T0275677

LONDON MATHEMATICAL SOCIETY LECTURE NOTE SERIES

Managing Editor: Professor M. Reid, Mathematics Institute,
University of Warwick, Coventry CV4 7AL, United Kingdom

The titles below are available from booksellers, or from Cambridge University Press at
http://www.cambridge.org/mathematics

343 Algebraic cycles and motives I, J. NAGEL & C. PETERS (eds)
344 Algebraic cycles and motives II, J. NAGEL & C. PETERS (eds)
345 Algebraic and analytic geometry, A. NEEMAN
346 Surveys in combinatorics 2007, A. HILTON & J. TALBOT (eds)
347 Surveys in contemporary mathematics, N. YOUNG & Y. CHOI (eds)
348 Transcendental dynamics and complex analysis, P.J. RIPPON & G.M. STALLARD (eds)
349 Model theory with applications to algebra and analysis I, Z. CHATZIDAKIS, D. MACPHERSON, A. PILLAY & A. WILKIE (eds)
350 Model theory with applications to algebra and analysis II, Z. CHATZIDAKIS, D. MACPHERSON, A. PILLAY & A. WILKIE (eds)
351 Finite von Neumann algebras and masas, A.M. SINCLAIR & R.R. SMITH
352 Number theory and polynomials, J. MCKEE & C. SMYTH (eds)
353 Trends in stochastic analysis, J. BLATH, P. MÖRTERS & M. SCHEUTZOW (eds)
354 Groups and analysis, K. TENT (ed)
355 Non-equilibrium statistical mechanics and turbulence, J. CARDY, G. FALKOVICH & K. GAWEDZKI
356 Elliptic curves and big Galois representations, D. DELBOURGO
357 Algebraic theory of differential equations, M.A.H. MACCALLUM & A.V. MIKHAILOV (eds)
358 Geometric and cohomological methods in group theory, M.R. BRIDSON, P.H. KROPHOLLER & I.J. LEARY (eds)
359 Moduli spaces and vector bundles, L. BRAMBILA-PAZ, S.B. BRADLOW, O. GARCÍA-PRADA & S. RAMANAN (eds)
360 Zariski geometries, B. ZILBER
361 Words: Notes on verbal width in groups, D. SEGAL
362 Differential tensor algebras and their module categories, R. BAUTISTA, L. SALMERÓN & R. ZUAZUA
363 Foundations of computational mathematics, Hong Kong 2008, F. CUCKER, A. PINKUS & M.J. TODD (eds)
364 Partial differential equations and fluid mechanics, J.C. ROBINSON & J.L. RODRIGO (eds)
365 Surveys in combinatorics 2009, S. HUCZYNSKA, J.D. MITCHELL & C.M. RONEY-DOUGAL (eds)
366 Highly oscillatory problems, B. ENGQUIST, A. FOKAS, E. HAIRER & A. ISERLES (eds)
367 Random matrices: High dimensional phenomena, G. BLOWER
368 Geometry of Riemann surfaces, F.P. GARDINER, G. GONZÁLEZ-DIEZ & C. KOUROUNIOTIS (eds)
369 Epidemics and rumours in complex networks, M. DRAIEF & L. MASSOULIÉ
370 Theory of p-adic distributions, S. ALBEVERIO, A.YU. KHRENNIKOV & V.M. SHELKOVICH
371 Conformal fractals, F. PRZYTYCKI & M. URBAŃSKI
372 Moonshine: The first quarter century and beyond, J. LEPOWSKY, J. MCKAY & M.P. TUITE (eds)
373 Smoothness, regularity and complete intersection, J. MAJADAS & A. G. RODICIO
374 Geometric analysis of hyperbolic differential equations: An introduction, S. ALINHAC
375 Triangulated categories, T. HOLM, P. JØRGENSEN & R. ROUQUIER (eds)
376 Permutation patterns, S. LINTON, N. RUŠKUC & V. VATTER (eds)
377 An introduction to Galois cohomology and its applications, G. BERHUY
378 Probability and mathematical genetics, N. H. BINGHAM & C. M. GOLDIE (eds)
379 Finite and algorithmic model theory, J. ESPARZA, C. MICHAUX & C. STEINHORN (eds)
380 Real and complex singularities, M. MANOEL, M.C. ROMERO FUSTER & C.T.C WALL (eds)
381 Symmetries and integrability of difference equations, D. LEVI, P. OLVER, Z. THOMOVA & P. WINTERNITZ (eds)
382 Forcing with random variables and proof complexity, J. KRAJÍČEK
383 Motivic integration and its interactions with model theory and non-Archimedean geometry I, R. CLUCKERS, J. NICAISE & J. SEBAG (eds)
384 Motivic integration and its interactions with model theory and non-Archimedean geometry II, R. CLUCKERS, J. NICAISE & J. SEBAG (eds)
385 Entropy of hidden Markov processes and connections to dynamical systems, B. MARCUS, K. PETERSEN & T. WEISSMAN (eds)
386 Independence-friendly logic, A.L. MANN, G. SANDU & M. SEVENSTER
387 Groups St Andrews 2009 in Bath I, C.M. CAMPBELL *et al* (eds)
388 Groups St Andrews 2009 in Bath II, C.M. CAMPBELL *et al* (eds)
389 Random fields on the sphere, D. MARINUCCI & G. PECCATI
390 Localization in periodic potentials, D.E. PELINOVSKY
391 Fusion systems in algebra and topology, M. ASCHBACHER, R. KESSAR & B. OLIVER
392 Surveys in combinatorics 2011, R. CHAPMAN (ed)
393 Non-abelian fundamental groups and Iwasawa theory, J. COATES *et al* (eds)
394 Variational problems in differential geometry, R. BIELAWSKI, K. HOUSTON & M. SPEIGHT (eds)
395 How groups grow, A. MANN
396 Arithmetic differential operators over the p-adic integers, C.C. RALPH & S.R. SIMANCA
397 Hyperbolic geometry and applications in quantum chaos and cosmology, J. BOLTE & F. STEINER (eds)
398 Mathematical models in contact mechanics, M. SOFONEA & A. MATEI
399 Circuit double cover of graphs, C.-Q. ZHANG

400 Dense sphere packings: a blueprint for formal proofs, T. HALES
401 A double Hall algebra approach to affine quantum Schur–Weyl theory, B. DENG, J. DU & Q. FU
402 Mathematical aspects of fluid mechanics, J.C. ROBINSON, J.L. RODRIGO & W. SADOWSKI (eds)
403 Foundations of computational mathematics, Budapest 2011, F. CUCKER, T. KRICK, A. PINKUS & A. SZANTO (eds)
404 Operator methods for boundary value problems, S. HASSI, H.S.V. DE SNOO & F.H. SZAFRANIEC (eds)
405 Torsors, étale homotopy and applications to rational points, A.N. SKOROBOGATOV (ed)
406 Appalachian set theory, J. CUMMINGS & E. SCHIMMERLING (eds)
407 The maximal subgroups of the low-dimensional finite classical groups, J.N. BRAY, D.F. HOLT & C.M. RONEY-DOUGAL
408 Complexity science: the Warwick master's course, R. BALL, V. KOLOKOLTSOV & R.S. MACKAY (eds)
409 Surveys in combinatorics 2013, S.R. BLACKBURN, S. GERKE & M. WILDON (eds)
410 Representation theory and harmonic analysis of wreath products of finite groups, T. CECCHERINI-SILBERSTEIN, F. SCARABOTTI & F. TOLLI
411 Moduli spaces, L. BRAMBILA-PAZ, O. GARCÍA-PRADA, P. NEWSTEAD & R.P. THOMAS (eds)
412 Automorphisms and equivalence relations in topological dynamics, D.B. ELLIS & R. ELLIS
413 Optimal transportation, Y. OLLIVIER, H. PAJOT & C. VILLANI (eds)
414 Automorphic forms and Galois representations I, F. DIAMOND, P.L. KASSAEI & M. KIM (eds)
415 Automorphic forms and Galois representations II, F. DIAMOND, P.L. KASSAEI & M. KIM (eds)
416 Reversibility in dynamics and group theory, A.G. O'FARRELL & I. SHORT
417 Recent advances in algebraic geometry, C.D. HACON, M. MUSTAŢĂ & M. POPA (eds)
418 The Bloch–Kato conjecture for the Riemann zeta function, J. COATES, A. RAGHURAM, A. SAIKIA & R. SUJATHA (eds)
419 The Cauchy problem for non-Lipschitz semi-linear parabolic partial differential equations, J.C. MEYER & D.J. NEEDHAM
420 Arithmetic and geometry, L. DIEULEFAIT *et al* (eds)
421 O-minimality and Diophantine geometry, G.O. JONES & A.J. WILKIE (eds)
422 Groups St Andrews 2013, C.M. CAMPBELL *et al* (eds)
423 Inequalities for graph eigenvalues, Z. STANIĆ
424 Surveys in combinatorics 2015, A. CZUMAJ *et al* (eds)
425 Geometry, topology and dynamics in negative curvature, C.S. ARAVINDA, F.T. FARRELL & J.-F. LAFONT (eds)
426 Lectures on the theory of water waves, T. BRIDGES, M. GROVES & D. NICHOLLS (eds)
427 Recent advances in Hodge theory, M. KERR & G. PEARLSTEIN (eds)
428 Geometry in a Fréchet context, C.T.J. DODSON, G. GALANIS & E. VASSILIOU
429 Sheaves and functions modulo p, L. TAELMAN
430 Recent progress in the theory of the Euler and Navier–Stokes equations, J.C. ROBINSON, J.L. RODRIGO, W. SADOWSKI & A. VIDAL-LÓPEZ (eds)
431 Harmonic and subharmonic function theory on the real hyperbolic ball, M. STOLL
432 Topics in graph automorphisms and reconstruction (2nd Edition), J. LAURI & R. SCAPELLATO
433 Regular and irregular holonomic D-modules, M. KASHIWARA & P. SCHAPIRA
434 Analytic semigroups and semilinear initial boundary value problems (2nd Edition), K. TAIRA
435 Graded rings and graded Grothendieck groups, R. HAZRAT
436 Groups, graphs and random walks, T. CECCHERINI-SILBERSTEIN, M. SALVATORI & E. SAVA-HUSS (eds)
437 Dynamics and analytic number theory, D. BADZIAHIN, A. GORODNIK & N. PEYERIMHOFF (eds)
438 Random walks and heat kernels on graphs, M.T. BARLOW
439 Evolution equations, K. AMMARI & S. GERBI (eds)
440 Surveys in combinatorics 2017, A. CLAESSON *et al* (eds)
441 Polynomials and the mod 2 Steenrod algebra I, G. WALKER & R.M.W. WOOD
442 Polynomials and the mod 2 Steenrod algebra II, G. WALKER & R.M.W. WOOD
443 Asymptotic analysis in general relativity, T. DAUDÉ, D. HÄFNER & J.-P. NICOLAS (eds)
444 Geometric and cohomological group theory, P.H. KROPHOLLER, I.J. LEARY, C. MARTÍNEZ-PÉREZ & B.E.A. NUCINKIS (eds)
445 Introduction to hidden semi-Markov models, J. VAN DER HOEK & R.J. ELLIOTT
446 Advances in two-dimensional homotopy and combinatorial group theory, W. METZLER & S. ROSEBROCK (eds)
447 New directions in locally compact groups, P.-E. CAPRACE & N. MONOD (eds)
448 Synthetic differential topology, M.C. BUNGE, F. GAGO & A.M. SAN LUIS
449 Permutation groups and cartesian decompositions, C.E. PRAEGER & C. SCHNEIDER
450 Partial differential equations arising from physics and geometry, M. BEN AYED *et al* (eds)
451 Topological methods in group theory, N. BROADDUS, M. DAVIS, J.-F. LAFONT & I.J. ORTIZ (eds)

London Mathematical Society Lecture Note Series: 451

Topological Methods in Group Theory

Edited by

N. BROADDUS
Ohio State University

M. DAVIS
Ohio State University

J.-F. LAFONT
Ohio State University

I.J. ORTIZ
Miami University

CAMBRIDGE
UNIVERSITY PRESS

University Printing House, Cambridge CB2 8BS, United Kingdom

One Liberty Plaza, 20th Floor, New York, NY 10006, USA

477 Williamstown Road, Port Melbourne, VIC 3207, Australia

314-321, 3rd Floor, Plot 3, Splendor Forum, Jasola District Centre, New Delhi - 110025, India

79 Anson Road, #06-04/06, Singapore 079906

Cambridge University Press is part of the University of Cambridge.

It furthers the University's mission by disseminating knowledge in the pursuit of education, learning and research at the highest international levels of excellence.

www.cambridge.org
Information on this title: www.cambridge.org/9781108437622
DOI: 10.1017/9781108526203

First published 2018

A catalogue record for this publication is available from the British Library

ISBN 978-1-108-43762-2 Paperback

Contents

Contributors

Antolín, Yago
Departamento de Matemáticas,
Universidad Autónoma de Madrid and Instituto de Ciencias Matemáticas
CSIC-UAM-UC3M-UCM ,
28049 Madrid, SPAIN
E-mail: yago.anpi@gmail.com

Berns-Zieve, Rose
Department of Mathematics and Statistics,
University of Massachusetts Amherst,
Amherst, MA 01003-9305, U.S.A.
E-mail: rbernszieve@verizon.net

Bieri, Robert
Fachbereich Mathematik,
Johann Wolfgang Goethe-Universität Frankfurt,
D-60054 Frankfurt am Main, GERMANY
E-mail: bieri@math.uni-frankfurt.de, rbieri@math.binghamton.edu

Bleak, Collin
School of Mathematics and Statistics,
University of St. Andrews,
St. Andrews, Fife, KY16 9SS, U.K.
E-mail: loeksnokes@gmail.com

Dicks, Warren
Departament de Matemàtiques,
Universitat Autònoma de Barcelona,
E-08193 Bellaterra (Barcelona), SPAIN
E-mail: dicks@mat.uab.cat

Farley, Daniel
Department of Mathematics,
Miami University
Oxford, OH 45056, U.S.A.
E-mail: farleyds@muohio.edu

Fry, Dana
Department of Mathematics,
University of Oregon,
Eugene, OR 97403-1222, U.S.A.
E-mail: dfry@uoregon.edu

Geoghegan, Ross
Department of Mathematical Sciences,
Binghamton University (SUNY),
Binghamton, NY 13902-6000, U.S.A.
E-mail: ross@math.binghamton.edu

Gillings, Johnny
Department of Mathematics,
North Carolina State University,
Raleigh, NC 27695, U.S.A.
E-mail: jrgillin@ncsu.edu

Guilbault, Craig R.
Department of Mathematical Sciences
University of Wisconsin-Milwaukee,
Milwaukee, WI 53201, U.S.A.
E-mail: craigg@uwm.edu

Hoganson, Hannah
Department of Mathematics,
University of Utah,
Salt Lake City, UT 84112-0090, U.S.A.
E-mail: hoganson@math.utah.edu

Jones, Keith
Department of Mathematics, Computer Science & Statistics,
State University of New York, College at Oneonta
Oneonta, NY 13820, U.S.A.
E-mail: keith.jones@oneonta.edu

Juschenko, Kate
Department of Mathematics,
Northwestern University,
Evanston, IL 60208, U.S.A.
E-mail: kate.juschenko@gmail.com

Kelsey, Gregory A.
Department of Mathematics,
Bellarmine University
Louisville, KY 40205, U.S.A.
E-mail: gkelsey@bellarmine.edu

Mathews, Heather
Department of Statistical Science,
Duke University,
Durham, NC 27708-0251, U.S.A.
E-mail: heather.mathews@duke.edu

Minemyer, Barry
Department of Mathematical and Digital Sciences,
Bloomsburg University,
Bloomsburg, PA 17815-1301, U.S.A.
E-mail: bminemyer@bloomu.edu

Nicas, Andrew
Department of Mathematics and Statistics,
McMaster University,
Hamilton, Ontario, L8S 4K1, CANADA
E-mail: nicas@mcmaster.ca

Qi, Dongwen
Department of Mathematics, Georgia Southwestern State University
Americus, GA 31709, U.S.A.
E-mail: dimayun@hotmail.com

Rosenthal, David
Department of Mathematics and Computer Science,
St. John's University,
Queens, NY 11439, U.S.A.
E-mail: rosenthd@stjohns.edu

Šunić, Zoran
Department of Mathematics,
Hofstra University,
Hempstead, NY 11549, U.S.A.
E-mail: zoran.sunic@hofstra.edu

Witzel, Stefan
Department of Mathematics,
Bielefeld University,
PO Box 100131, 33501 Bielefeld, GERMANY
E-mail: switzel@math.uni-bielefeld.de

Zaremsky, Matthew C. B.
Department of Mathematics and Statistics,
University at Albany (SUNY),
Albany, NY 12222, U.S.A.
E-mail: mzaremsky@albany.edu

Preface

The field of geometric group theory is built around one overarching philosophy – to understand a group G by understanding a space X on which it acts. This prompts a rich back and forth, between properties of the group, and properties of the space. Since a given group can act on many different spaces, a major theme is to try and find the "best" space on which G acts. This naturally leads to a strong interplay between geometric group theory and topology.

One of the early proponents of this method was Ross Geoghegan. In a famous 1984 paper with Ken Brown, Geoghegan showed that Thompson's group F provided an example of a finitely presented, torsion-free, FP_∞ group that has infinite cohomological dimension. After working in other topological areas, Geoghegan returned to this topic in the mid-1990s. Around this time, he started a productive collaboration with Robert Bieri, in which they explored the controlled connectivity properties of horoballs in non-positively curved spaces – a project which is still ongoing. In 2008, Geoghegan wrote an important textbook with the same title as these proceedings. It provided an introduction to the use of topological techniques in the study of infinite groups.

In June 2014, the mathematics department at Ohio State hosted the conference "Topological Methods in Group Theory", in honor of Ross Geoghegan's 70th birthday. The week-long conference featured 19 plenary talks, and 22 short talks, on a variety of topics in geometric group theory. The present volume contains the proceedings of the conference. The articles in this volume cover a wide cross-section of topics in geometric group theory, including left-orderable groups, groups defined by automata, connectivity properties and Σ-invariants of groups, amenability and non-amenability problems, and boundaries of certain groups. It also covers topics which are more geometric or topological in nature, such as the geometry of simplices, decomposition complexity of certain groups, and problems in shape theory. We hope that, through this volume, the reader will obtain a taste of the rich mathematics presented at the conference.

N. Broaddus
Department of Mathematics, The Ohio State University
231 West 18th Avenue, Columbus, OH 43210-1174, U.S.A.
e-mail: broaddus.9@osu.edu

M. W. Davis
Department of Mathematics, The Ohio State University
231 West 18th Avenue, Columbus, OH 43210-1174, U.S.A.
e-mail: davis.12@osu.edu

J.-F. Lafont
Department of Mathematics, The Ohio State University
231 West 18th Avenue, Columbus, OH 43210-1174, U.S.A.
e-mail: lafont.1@osu.edu

I. J. Ortiz
Department of Mathematics, Miami University
Oxford, OH 45056, U.S.A.
e-mail: ortizi@muohio.edu

Acknowledgments

We, as part of the organizing committee, thoroughly enjoyed working on this event. We would like to take this opportunity to thank all the people who helped make the conference, and these proceedings, possible. Specifically, we would like to thank the speakers and participants for a very enjoyable conference. We would also like to thank the contributors, as well as the anonymous referees, for all their hard work on the present volume.

Finally, we would like to acknowledge the financial support of the NSF (under grant DMS-1441592) and the Ohio State University's Mathematical Research Institute. Their generous support was instrumental in making this event, and these proceedings, possible.

1
Left Relatively Convex Subgroups

Yago Antolín, Warren Dicks, and Zoran Šunić

Abstract

Let G be a group and H be a subgroup of G. We say that H is left relatively convex in G if the left G-set G/H has at least one G-invariant order; when G is left orderable, this holds if and only if H is convex in G under some left ordering of G.

We give a criterion for H to be left relatively convex in G that generalizes a famous theorem of Burns and Hale and has essentially the same proof. We show that all maximal cyclic subgroups are left relatively convex in free groups, in right-angled Artin groups, and in surface groups that are not the Klein-bottle group. The free-group case extends a result of Duncan and Howie.

More generally, every maximal m-generated subgroup in a free group is left relatively convex. The same result is valid, with some exceptions, for compact surface groups. Maximal m-generated abelian subgroups in right-angled Artin groups are left relatively convex.

If G is left orderable, then each free factor of G is left relatively convex in G. More generally, for any graph of groups, if each edge group is left relatively convex in each of its vertex groups, then each vertex group is left relatively convex in the fundamental group; this generalizes a result of Chiswell.

All maximal cyclic subgroups in locally residually torsion-free nilpotent groups are left relatively convex.

1.1 Outline

Notation 1.1 Throughout this chapter, let G be a multiplicative group, and G_0 be a subgroup of G. For $x, y \in G$, $[x,y] := x^{-1}y^{-1}xy$, $x^y := y^{-1}xy$, and ${}^y x := yxy^{-1}$. For any subset X of G, we denote by $X^{\pm 1} := X \cup X^{-1}$, by $\langle X \rangle$ the subgroup of G generated by X, by $\langle X^G \rangle$ the normal subgroup of G generated by X, and let $G/\lhd X \rhd := G/\langle X^G \rangle$. When we write $A \subseteq B$ we mean that A is a subset of B, and when we write $A \subset B$ we mean that A is a proper subset of B.

In Section 1.2, we collect together some facts, several of which first arose in the proof of Theorem 28 of [5]. If G is left orderable, Bergman calls G_0 'left relatively convex in G' if G_0 is convex in G under some left ordering of G, or, equivalently, the left G-set G/G_0 has some G-invariant order. Broadening the scope of his terminology, we shall say that G_0 is *left relatively convex in* G if the left G-set G/G_0 has some G-invariant order, even if G is not left orderable.

We give a criterion for G_0 to be left relatively convex in G that generalizes a famous theorem of Burns and Hale [7] and has essentially the same proof. We deduce that if each noncyclic, finitely generated subgroup of G maps onto $\mathbb{Z}^2$, then each maximal cyclic subgroup of G is left relatively convex in G. Thus, if F is a free group and C is a maximal cyclic subgroup of F, then F/C has an F-invariant order; this extends the result of Duncan and Howie [15] that a certain finite subset of F/C has an order that is respected by the partial F-action. Louder and Wilton [21] used the Duncan–Howie order to prove Wise's conjecture that, for subgroups H and K of a free group F, if H or K is a maximal cyclic subgroup of F, then $\sum_{HxK \in H\backslash F/K} \operatorname{rank}(H^x \cap K) \leq \operatorname{rank}(H)\operatorname{rank}(K)$. They also gave a simple proof of the existence of a Duncan–Howie order; translating their argument from topological to algebraic language led us to the order on F/C. More generally, we introduce the concept of n-indicability and use it to show that each maximal m-generated subgroup of a free group is left relatively convex.

In Section 1.3, we find that the main result of [13] implies that, for any graph of groups, if each edge group is left relatively convex in each of its vertex groups, then each vertex group is left relatively convex in the fundamental group. This generalizes a result of Chiswell [8]. In particular, in a left-orderable group, each free factor is left relatively convex.

One says that G is *discretely left orderable* if some infinite (maximal) cyclic subgroup of G is left relatively convex in G. Many examples of such groups are given in [20]; for instance, it is seen that among free groups, braid groups, surface groups, and right-angled Artin groups, all the infinite ones are discretely left orderable. In Section 1.4, we show that all maximal cyclic subgroups are left relatively convex in right-angled Artin groups and in compact surface groups that are not the Klein-bottle group. More generally, we show that, with some exceptions, each maximal m-generated subgroup of a compact surface group is left relatively convex, and each maximal m-generated abelian subgroup of a right-angled Artin group is left relatively convex.

At the end, in Section 1.5, we show that all maximal cyclic subgroups in locally residually torsion-free nilpotent groups are left relatively convex.

1.2 Left Relatively Convex Subgroups

Definition 1.2 Let X be a set and $\mathcal{R}$ be a binary relation on X; thus, $\mathcal{R}$ is a subset of $X \times X$, and '$x\mathcal{R}y$' means '$(x, y) \in \mathcal{R}$'. We say that $\mathcal{R}$ is *transitive* when, for all x, y, $z \in X$, if $x\mathcal{R}y$ and $y\mathcal{R}z$, then $x\mathcal{R}z$, and here we write $x\mathcal{R}y\mathcal{R}z$ and say that *y fits between x and z with respect to $\mathcal{R}$*. We say that $\mathcal{R}$ is *trichotomous* when, for all x, $y \in X$, exactly one of $x\mathcal{R}y$, $x = y$, and $y\mathcal{R}x$ holds, and here we say that the *sign* of the triple $(x, \mathcal{R}, y)$, denoted $\text{sign}(x, \mathcal{R}, y)$, is 1, 0, or -1, respectively. A transitive, trichotomous binary relation is called an *order*. For any order $<$ on X, a subset Y of X is said to be *convex in X with respect to* $<$ if no element of $X - Y$ fits between two elements of Y with respect to $<$.

Now suppose that X is a left G-set. The diagonal left G-action on $X \times X$ gives a left G-action on the set of binary relations on X. By a *binary G-relation on X* we mean a G-invariant binary relation on X, and by a *G-order on X* we mean a G-invariant order on X. If there exists at least one G-order on X, we say that X is *G-orderable*. If X is endowed with a G-order, we say that X is *G-ordered*. When X is G with the left multiplication action, we replace 'G-' with 'left', and write *left order*, *left orderable*, or *left ordered*, the latter two being hyphenated when they premodify a noun.

Analogous terminology applies for right G-sets.

Definition 1.3 For $K \leq H \leq G$, we recall two mutually inverse operations. Let x, $y \in G$.

If $<$ is a G-order on G/K with respect to which H/K is convex in G/K, then we define an H-order $<_{\text{bottom}}$ on H/K and a G-order $<_{\text{top}}$ on G/H as follows. We take $<_{\text{bottom}}$ to be the restriction of $<$ to H/K. We define $xH <_{\text{top}} yH$ to mean $(\forall h_1, h_2 \in H)(xh_1K < yh_2K)$. This relation is trichotomous since $xH <_{\text{top}} yH$ holds if and only if we have $(xH \neq yH) \wedge (xK < yK)$; the former clearly implies the latter, and, when the latter holds, $K < x^{-1}yK$, and then, by the convexity of H/K in G/K, $h_1K < x^{-1}yK$, and then $y^{-1}xh_1K < K$, $y^{-1}xh_1K < h_2K$, and $xh_1K < yh_2K$. Thus, $<_{\text{top}}$ is a G-order on G/H.

Conversely, if $<_{\text{bottom}}$ is an H-order on H/K and $<_{\text{top}}$ is a G-order on G/H, we now define a G-order $<$ on G/K with respect to which H/K is convex in G/K. We define $xK < yK$ to mean

$$(xH <_{\text{top}} yH) \ \vee \ \big((xH = yH) \wedge (K <_{\text{bottom}} x^{-1}yK)\big).$$

It is clear that $<$ is a well-defined G-order on G/K. Now suppose that $xK \in (G/K) - (H/K)$. Then $xH \neq H$. If $xH <_{\text{top}} H$, then $xK < hK$, for all $h \in H$, and similarly if $H <_{\text{top}} xH$. Thus, H/K is convex in G/K with respect to $<$.

In particular, G/K has some G-order with respect to which H/K is convex in G/K if and only if H/K is H-orderable and G/H is G-orderable. Taking $K = \{1\}$ and $H = G_0$, we find that the following are equivalent, as seen in the proof of Theorem 28 (vii)⇔(viii) of [5]:

(1.3.1) G has some left order with respect to which G_0 is convex in G,
(1.3.2) G_0 is left orderable, and G/G_0 is G-orderable,
(1.3.3) G is left orderable, and G/G_0 is G-orderable.

This motivates the terminology introduced in the following definition, which presents an analysis similar to one given by Bergman in the proof of Theorem 28 in [5]. Unlike Bergman, we do not require that the group G is left-ordered.

Definition 1.4 Let $\text{Ssg}(G)$ denote the set of all the subsemigroups of G, that is, subsets of G closed under the multiplication. We say that the subgroup G_0 *of* G *is left relatively convex in* G when any of the following equivalent conditions hold:

(1.4.1) the left G-set G/G_0 is G-orderable,
(1.4.2) the right G-set $G_0\backslash G$ is G-orderable,
(1.4.3) there exists some $G_+ \in \text{Ssg}(G)$ such that $G_+^{\pm 1} = G - G_0$; in this event, $G_+ \cap G_+^{-1} = \emptyset$ and $G_0G_+ = G_+G_0 = G_0G_+G_0 = G_+$,

(1.4.4) for each finite subset X of $G{-}G_0$, there exists $S \in \mathrm{Ssg}(G)$ such that $X \subseteq S^{\pm 1} \subseteq G{-}G_0$.

We then say also that G_0 *is a left relatively convex subgroup of* G. One may also use 'right' in place of 'left'.

Proof of equivalence (1.4.1)⇒(1.4.3). Let $<$ be a G-order on G/G_0, and set

$$G_+ := \{x \in G \mid G_0 < xG_0\};$$

then $G_+^{-1} = \{x \in G \mid G_0 < x^{-1}G_0\} = \{x \in G \mid xG_0 < G_0\}$ and $G_0 = \{x \in G \mid G_0 = xG_0\}$. Hence, $G_+^{\pm 1} = G{-}G_0$. If x, $y \in G_+$, then $G_0 < xG_0$, $G_0 < yG_0$ and $G_0 < xG_0 < xyG_0$; thus $xy \in G_+$. Hence, $G_+ \in \mathrm{Ssg}(G)$.

Now consider any $G_+ \in \mathrm{Ssg}(G)$ such that $G_+^{\pm 1} = G{-}G_0$. Then $G_+ \cap G_+^{-1} = \emptyset$, since G_+ is a subsemigroup which does not contain 1. Also, $G_0G_+ \cap G_0 = \emptyset$, since $G_+ \cap G_0^{-1}G_0 = \emptyset$, while $G_0G_+ \cap G_+^{-1} = \emptyset$, since $G_0 \cap G_+^{-1}G_+^{-1} = \emptyset$. Thus $G_0G_+ \subseteq G_+$, and equality must hold. Similarly, $G_+G_0 = G_+$.

(1.4.3)⇒(1.4.1). Let x, y, $z \in G$. We define $xG_0 < yG_0$ to mean that $(xG_0)^{-1}(yG_0) \subseteq G_+$, or, equivalently, that $x^{-1}y \in G_+$. Then $<$ is a well-defined binary G-relation on G/G_0. Since $x^{-1}y$ belongs to exactly one of G_+, G_0, and G_+^{-1}, we see that $<$ is trichotomous. If $xG_0 < yG_0$ and $yG_0 < zG_0$, then G_+ contains $x^{-1}y$, $y^{-1}z$, and their product, which shows that $xG_0 < zG_0$. Thus $<$ is a G-order on G/G_0.

(1.4.2)⇔(1.4.3) is the left-right dual of (1.4.1)⇔(1.4.3).

(1.4.3)⇒(1.4.4) with $S = G_+$.

(1.4.4)⇒(1.4.3). Bergman [5] observes that an implication of this type follows easily from the Compactness Theorem of Model Theory; here, one could equally well use the quasi-compactness of $\{-1,1\}^{G-G_0}$, which holds by a famous theorem of Tychonoff [27]. The case of this implication where $G_0 = \{1\}$ was first stated by Conrad [9], who gave a short argument designed to be read in conjunction with a short argument of Ohnishi [25]. Let us show that a streamlined form of the Conrad–Ohnishi proof gives the general case comparatively easily.

Let 2^{G-G_0} denote the set of all subsets of $G{-}G_0$. For each $W \in 2^{G-G_0}$, let $\mathrm{Fin}(W)$ denote the set of finite subsets of W, and $\langle\langle W\rangle\rangle$ denote the subsemigroup of G generated by W. For each $\varphi \in \{-1,1\}^{G-G_0}$ and $x \in G{-}G_0$, set $\tilde{\varphi}(x) := x^{\varphi(x)} \in \{x, x^{-1}\}$. Set

$$\mathfrak{W} := \Big\{ W \in 2^{G-G_0} \mid (\forall\, W' \in \mathrm{Fin}(W))\ (\forall X \in \mathrm{Fin}(G{-}G_0)) \\ (\exists \varphi \in \{-1,1\}^{G-G_0})\Big(G_0 \cap \langle\langle\, W' \cup \tilde{\varphi}(X)\,\rangle\rangle = \emptyset\Big)\Big\}.$$

It is not difficult to see that (1.4.4) says precisely that $\emptyset \in \mathfrak{W}$. Also, it is clear that

$$(\forall\, W \in 2^{G-G_0})\Big((W \in \mathfrak{W}) \Leftrightarrow (\mathrm{Fin}(W) \subseteq \mathfrak{W})\Big).$$

It follows that $\mathfrak{W}$ is closed in 2^{G-G_0} under the operation of taking unions of chains. By Zorn's Lemma, there exists some maximal element W of $\mathfrak{W}$.

We shall prove that $\langle\langle W\rangle\rangle^{\pm 1} = G{-}G_0$, and thus (1.4.3) holds. By taking $X = \emptyset$ in the definition of '$W \in \mathfrak{W}$', we see that $\langle\langle W\rangle\rangle \subseteq G{-}G_0$, and thus $W^{\pm 1} \subseteq \langle\langle W\rangle\rangle^{\pm 1} \subseteq G{-}G_0$. It remains to show that $G{-}G_0 \subseteq W^{\pm 1}$. Since W is maximal in $\mathfrak{W}$, it suffices to show that

$$(\forall x \in G{-}G_0)\big((W \cup \{x\} \in \mathfrak{W}) \vee (W \cup \{x^{-1}\} \in \mathfrak{W})\big).$$

Suppose then $W \cup \{x\} \notin \mathfrak{W}$. Thus, we may fix a $W_x \in \mathrm{Fin}(W)$ and an $X_x \in \mathrm{Fin}(G{-}G_0)$ such that

$$(\forall \varphi \in \{-1,1\}^{G-G_0})\ \Big(G_0\ \cap\ \langle\langle\ W_x \cup \{x\} \cup \tilde{\varphi}(X_x)\ \rangle\rangle \neq \emptyset\Big).$$

Let $W' \in \mathrm{Fin}(W)$ and $X \in \mathrm{Fin}(G{-}G_0)$. As $W \in \mathfrak{W}$, there exists a function $\varphi \in \{-1,1\}^{G-G_0}$ such that

$$G_0 \cap \langle\langle\ W_x \cup W' \cup \tilde{\varphi}(\{x\} \cup X_x \cup X)\ \rangle\rangle = \emptyset.$$

Clearly, $\tilde{\varphi}(x) \neq x$. Thus, $\tilde{\varphi}(x) = x^{-1}$ and

$$G_0 \cap \langle\langle\ W' \cup \{x^{-1}\} \cup \tilde{\varphi}(X)\ \rangle\rangle = \emptyset.$$

This shows that $W \cup \{x^{-1}\} \in \mathfrak{W}$, as desired. □

The Burns–Hale theorem [7, Theorem 2] says that if each nontrivial, finitely generated subgroup of G maps onto some nontrivial, left-orderable group, then G is left orderable. The following result, using a streamlined version of their proof, generalizes the Burns–Hale theorem in two ways. Namely, the scope is increased by stating the result for an arbitrary subgroup G_0 (in their case G_0 is trivial) and by imposing a weaker condition (in their case $\langle X\rangle$ is required to map onto a left-orderable group).

Theorem 1.5 *If, for each nonempty, finite subset X of $G - G_0$, there exists a proper, left relatively convex subgroup of $\langle X\rangle$ that includes $\langle X\rangle \cap G_0$, then G_0 is left relatively convex in G.*

Proof For each finite subset X of $G{-}G_0$, we shall construct an element $S_X \in \mathrm{Ssg}(\langle X\rangle)$ such that $X \subseteq S_X^{\pm 1} \subseteq G{-}G_0$, and then (1.4.4) will hold. We set $S_\emptyset := \emptyset$. We now assume that $X \neq \emptyset$. Let us write $H := \langle X\rangle$. By hypothesis, we have an H_0 such that $H \cap G_0 \leq H_0 < H$ and H_0 is left relatively convex in H. Notice that $H{-}H_0 \subseteq H{-}(H \cap G_0) \subseteq G{-}G_0$ and $X \cap H_0 \subset X$, since $X \not\subseteq H_0$. By induction on $|X|$, we can find an $S_{X\cap H_0} \in \mathrm{Ssg}(\langle X \cap H_0\rangle)$ such that $X \cap H_0 \subseteq S_{X\cap H_0}^{\pm 1} \subseteq G{-}G_0$. By (1.4.3), since H_0 is left relatively convex in H, we have an $H_+ \in \mathrm{Ssg}(H)$ such that $H_0H_+H_0 = H_+$ and $H_+^{\pm 1} = H - H_0$. We set $S_X := S_{X \cap H_0} \cup H_+$. Then $S_X \in \mathrm{Ssg}(H)$, since $S_{X \cap H_0} \subseteq H_0$ and $H_0H_+H_0 = H_+$. Also,

$$X = (X \cap H_0) \cup (X - H_0) \subseteq S_{X \cap H_0}^{\pm 1} \cup (H - H_0) = S_X^{\pm 1} \subseteq G - G_0. \quad \square$$

Remark Theorem 1.5 has a variety of corollaries. For example, for any subset X of G, we have a sequence of successively weaker conditions: $\langle X \cup G_0 \rangle / \lhd G_0 \rhd$ maps onto $\mathbb{Z}$; $\langle X \cup G_0 \rangle / \lhd G_0 \rhd$ maps onto a nontrivial, left-orderable group; there exists a proper, left relatively convex subgroup of $\langle X \cup G_0 \rangle$ that includes G_0; and, there exists a proper, left relatively convex subgroup of $\langle X \rangle$ that includes $\langle X \rangle \cap G_0$. The last implication follows from the following fact. If A and B are subgroups of G and A is left relatively convex in G, then $A \cap B$ is left relatively convex in B.

Definition 1.6 A group G is said to be *n-indicable*, where n is a positive integer, if it can be generated by fewer than n elements or it admits a surjective homomorphism onto $\mathbb{Z}^n$.

A group G is *locally n-indicable* if every finitely generated subgroup of G is n-indicable.

Note that some authors require in the definition of indicability that G admits a surjective homomorphism onto $\mathbb{Z}$, while here 1-indicable means that G is trivial or maps onto $\mathbb{Z}$, 2-indicable means that G is cyclic or maps onto $\mathbb{Z}^2$, and so on.

Example 1.7 Free abelian groups of any rank and free groups of any rank are locally n-indicable for every n.

The notion of n-indicability is related to left relative convexity through the following corollary of Theorem 1.5.

Corollary 1.8 *Let $n \geq 2$. If G is a locally n-indicable group then each maximal $(n-1)$-generated subgroup of G is left relatively convex in G. In particular, in a free group, each maximal cyclic subgroup is left relatively convex.*

Proof If the subgroup G_0 is a maximal $(n-1)$-generated subgroup of G, then, for any nonempty, finite subset X of $G - G_0$, $\langle X \cup G_0 \rangle$ maps onto $\mathbb{Z}^n$, and $\langle X \cup G_0 \rangle / \lhd G_0 \rhd$ maps onto $\mathbb{Z}$. $\square$

The idea of Corollary 1.8 can be used to show that certain maximal κ-generated abelian subgroups are left relatively convex, where κ is some cardinal.

Definition 1.9 A group G is *nasmof* if it is torsion-free and every nonabelian subgroup of G admits a surjective homomorphism onto $\mathbb{Z}*\mathbb{Z}$.

Example 1.10 The class of nasmof groups contains free and free abelian groups and it is closed under taking subgroups and direct products. Residually nasmof groups are nasmof, and in particular residually free groups are nasmof. Every nasmof group G is 2-locally indicable, and by Corollary 1.8, maximal cyclic subgroups are left relatively convex.

Corollary 1.11 *Let κ be a cardinal. If G is a nasmof group then each maximal κ-generated abelian subgroup of G is left relatively convex in G.*

In particular, in a residually free group, each maximal κ-generated abelian subgroup is left relatively convex.

Proof Let G_0 be a maximal κ-generated abelian subgroup of G and X a nonempty finite subset of $G - G_0$. By maximality, if $\langle X \cup G_0 \rangle$ is abelian, then it is not κ-generated and κ must be a finite cardinal. In this case, $\langle X \cup G_0 \rangle$ is a finitely generated, torsion-free abelian group of rank greater than κ. If $\langle X \cup G_0 \rangle$ is nonabelian, then it maps onto $\mathbb{Z} * \mathbb{Z}$. In both cases, $\langle X \cup G_0 \rangle / \lhd G_0 \rhd$ maps onto $\mathbb{Z}$. □

1.3 Graphs of Groups

Definition 1.12 By a *graph*, we mean a quadruple (Γ, V, ι, τ) such that Γ is a set, V is a subset of Γ, and ι and τ are maps from $\Gamma - V$ to V. Here, we let Γ denote the graph as well as the set, and we write $\mathrm{V}\Gamma := V$ and $\mathrm{E}\Gamma := \Gamma - V$, called the *vertex-set* and *edge-set*, respectively. We then define *vertex*, *edge* $\iota e \xrightarrow{e} \tau e$, *inverse edge* $\tau e \xrightarrow{e^{-1}} \iota e$, *path*

$$(1.12.1) \qquad v_0 \xrightarrow{e_1^{\epsilon_1}} v_1 \xrightarrow{e_2^{\epsilon_2}} v_2 \xrightarrow{e_3^{\epsilon_3}} \cdots \xrightarrow{e_{n-2}^{\epsilon_{n-2}}} v_{n-2} \xrightarrow{e_{n-1}^{\epsilon_{n-1}}} v_{n-1} \xrightarrow{e_n^{\epsilon_n}} v_n, \; n \geq 0,$$

reduced path, and *connected graph* in the usual way. We say that Γ is a *tree* if $V \neq \emptyset$ and, for each $(v, w) \in V \times V$, there exists a unique reduced path from v to w. The *barycentric subdivision of Γ* is the graph $\Gamma^{(\prime)}$ such that $\mathrm{V}\Gamma^{(\prime)} = \Gamma$ and $\mathrm{E}\,\Gamma^{(\prime)} = \mathrm{E}\,\Gamma \times \{\iota, \tau\}$, with $e \xrightarrow{(e,\iota)} \iota e$ and $e \xrightarrow{(e,\tau)} \tau e$.

We say that Γ is a *left G-graph* if Γ is a left G-set, V is a G-subset of Γ, and ι and τ are G-maps. For $\gamma \in \Gamma$, we let G_γ denote the G-stabilizer of γ.

Let T be a tree. A *local order on T* is a family $(<_v \mid v \in \mathrm{V}T)$ such that, for each $v \in \mathrm{V}T$, $<_v$ is an order on $\mathrm{link}_T(v) := \{e \in \mathrm{E}T \mid v \in \{\iota e, \tau e\}\}$. By Theorem 3 of [13], for each local order $(<_v \mid v \in \mathrm{V}T)$ on T, there exists a unique order $<_T$ on $\mathrm{V}T$ such that, for each reduced T-path expressed as in (1.12.1),

$$\mathrm{sign}(v_0, <_T, v_n) = \mathrm{sign}\big(0,\ <_{\mathbb{Z}},\ \sum_{i=1}^{n} \epsilon_i + \sum_{i=1}^{n-1} \mathrm{sign}(e_i, <_{v_i}, e_{i+1})\big),$$

where the sign notation is as in Definition 1.2. We then call $<_T$ the *associated order*, $\sum_{i=1}^{n} \epsilon_i$ the *orientation-sum*, and $\sum_{i=1}^{n-1} \mathrm{sign}(e_i, <_{v_i}, e_{i+1})$ the *turn-sum*. If T is a left G-tree, then, for any G-invariant local order on T, the associated order on $\mathrm{V}T$ is easily seen to be a G-order.

Theorem 1.13 *Suppose that T is a left G-tree such that, for each T-edge e, G_e is left relatively convex in $G_{\iota e}$ and in $G_{\tau e}$. Then, for each $t \in T$, G_t is left relatively convex in G. If there exists some $t \in T$ such that G_t is left orderable, then G is left orderable. Moreover, if the input orders are given effectively, then the output orders are given effectively,*

Proof We choose one representative from each G-orbit in $\mathrm{V}T$. For each representative v_0, we choose an arbitrary order on the set of G_{v_0}-orbits $G_{v_0}\backslash \mathrm{link}_T(v_0)$, and, within each G_{v_0}-orbit, we choose one representative e_0 and a G_{v_0}-order on G_{v_0}/G_{e_0}, which exists by (1.4.1); since our G_{v_0}-orbit $G_{v_0}e_0$ may be identified with G_{v_0}/G_{e_0}, we then have a G_{v_0}-order on $G_{v_0}e_0$, and then on all of $\mathrm{link}_T(v_0)$ by our order on $G_{v_0}\backslash \mathrm{link}_T(v_0)$. We then use G-translates to obtain a G-invariant local order on T. This in turn gives the associated G-order on $\mathrm{V}T$ as in Definition 1.12. In particular, for each T-vertex v, we have G-orders on Gv and G/G_v. By (1.4.1), G_v is then left relatively convex in G. For each T-edge e, G_e is left relatively convex in $G_{\iota e}$ by hypothesis, and then G_e is left relatively convex in G by Definition 1.3. Thus, for each $t \in T$, G_t is left relatively convex in G.

By (1.3.2)⇒(1.3.3), if there exists some $t \in T$ such that G_t is left orderable, then G is left orderable. □

Example 1.14 Let F be a free group and X be a free-generating set of F. The left Cayley graph of F with respect to X is a left F-tree on which F acts freely. Thus, the fact that free groups are left orderable can be deduced from Theorem 1.13; see [13].

Bearing in mind that intersections of left relatively convex subgroups are left relatively convex, we can generalize the previous example to the case that a group acts freely on some orbit of n-tuples of elements of T.

Corollary 1.15 *Suppose that T is a left G-tree such that, for each T-edge e, G_e is left relatively convex in $G_{\iota e}$ and in $G_{\tau e}$. Suppose that there exists a finite subset S of T with $\cap_{s \in S} G_s = \{1\}$, then G is left orderable.*

Definition 1.16 By a *graph of groups* $(\mathfrak{G}, \Gamma)$, we mean a graph with vertex-set a family of groups $(\mathfrak{G}(v') \mid v' \in \mathrm{V}\,\Gamma^{(\prime)})$ and edge-set a family of injective group homomorphisms $(\mathfrak{G}(e) \xrightarrow{\mathfrak{G}(e')} \mathfrak{G}(v) \mid e \xrightarrow{e'} v \in \mathrm{E}\,\Gamma^{(\prime)})$, where Γ is a nonempty, connected graph and $\Gamma^{(\prime)}$ is its barycentric subdivision. For $\gamma \in \Gamma^{(\prime)}$, we call $\mathfrak{G}(\gamma)$ a *vertex group*, *edge group*, or *edge map* if γ belongs to $\mathrm{V}\Gamma$, $\mathrm{E}\Gamma$, or $\mathrm{E}\,\Gamma^{(\prime)}$, respectively. One may think of $(\mathfrak{G}, \Gamma)$ as a nonempty, connected graph, of groups and injective group homomorphisms, in which every vertex is either a sink, called a vertex group, or a source of valence two, called an edge group. We shall use the *fundamental group* and the *Bass–Serre tree* of $(\mathfrak{G}, \Gamma)$ as defined in [26] and [11].

Bass–Serre theory translates Theorem 1.13 into the following form.

Theorem 1.17 *Suppose that G is the fundamental group of a graph of groups $(\mathfrak{G}, \Gamma)$ such that the image of each edge map $\mathfrak{G}(e) \xrightarrow{\mathfrak{G}(e')} \mathfrak{G}(v)$ is left relatively convex in its vertex group, $\mathfrak{G}(v)$. Then each vertex group is left relatively convex in G. If some vertex group is left orderable, then G is left orderable. Moreover, if the input orders are given effectively, then the output orders are given effectively.* □

Remark Theorem 1.17 generalizes the result of Chiswell that a group is left orderable if it is the fundamental group of a graph of groups such that each vertex group is left ordered and each edge group is convex in each of its vertex groups; see Corollary 3.5 of [8].

The result of Chiswell is a consequence of Corollary 3.4 of [8], which shows that a group is left orderable if it is the fundamental group of a graph of groups such that each edge group is left orderable and each of its left orders extends to a left order on each of its vertex groups. (If, moreover, each edge group and vertex group is left ordered, and the maps from edge groups to vertex groups respect the orders, then the fundamental group has a left order such that the maps from the vertex groups to the fundamental group respect the orders.) This applies to the case of cyclic edge groups and left-orderable vertex groups.

Corollary 3.4 of [8] is, in turn, a consequence of Chiswell's necessary and sufficient conditions for the fundamental group of a graph of groups to be left orderable. As his proof involved ultraproducts, his orders were not constructed effectively.

Example 1.18 Let A and B be groups, C be a subgroup of A, and $x : C \to B$, $c \mapsto c^x$, be an injective homomorphism. The graph of groups $A \leftarrow C \to B$, where the maps are the inclusion map and x, has as fundamental group $A *_C B := A * B/\lhd\{c^{-1}\cdot c^x \mid c \in C\}\rhd$, called the *free product with amalgamation* with *vertex groups A and B*, *edge group C*, and *edge map x*. We then view A and B as subgroups of $A *_C B$. In particular, $c^x = c$.

If C is left relatively convex in each of A and B, then A and B are left relatively convex in $A *_C B$, by Theorem 1.17.

In detail, suppose that $G = A *_C B$, that $<_A$ is an A-order on A/C, and that $<_B$ is a B-order on B/C. The Bass–Serre left G-tree T for $A \leftarrow C \to B$ has vertex-set $G/A \mathbin{\dot\cup} G/B$ (where $\dot\cup$ denotes the disjoint union) and edge-set G/C, with $gA \xrightarrow{gC} gB$. Then $<_A$ and $<_B$ determine a G-invariant local order on T, and we have the associated G-order $<_T$ on $\mathrm{V}\,T$, as in Definition 1.12. Let us describe the G-order $<_T$ on G/A. Consider any $gA \in G/A$, and write $gA = a_1b_1a_2b_2\cdots a_nb_nA$, $n \geq 0$, where $a_1 \in A$, $a_2, \ldots, a_n \in A{-}C$, $b_1, b_2, \ldots, b_n \in B{-}C$. We then have a reduced T-path

$$A \xrightarrow{a_1C} a_1B \xrightarrow{(a_1b_1C)^{-1}} a_1b_1A \xrightarrow{a_1b_1a_2C} a_1b_1a_2B \xrightarrow{(a_1b_1a_2b_2C)^{-1}} \cdots \xrightarrow{a_1b_1a_2b_2\cdots a_nC} a_1b_1a_2b_2\cdots a_nB \xrightarrow{(a_1b_1a_2b_2\cdots a_nb_nC)^{-1}} a_1b_1a_2b_2\cdots a_nb_nA = gA.$$

The orientation-sum equals zero, and we have only the turn-sum, which simplifies by the G-invariance of the local order to give

$$\mathrm{sign}(A, <_T, gA) = \mathrm{sign}\Big(0,\ <_{\mathbb{Z}},\ \sum_{i=1}^{n} \mathrm{sign}(C, <_B, b_iC) + \sum_{i=2}^{n} \mathrm{sign}(C, <_A, a_iC)\Big).$$

We record the case where $C = \{1\}$.

Corollary 1.19 *In a left-orderable group, every free factor is left relatively convex.* □

Example 1.20 In a free group, every free factor is left relatively convex, by Example 1.14.

Example 1.21 Suppose that A and B are free groups, or, more generally, groups all of whose maximal cyclic subgroups are left relatively convex; see Corollary 1.8. If C is a maximal cyclic subgroup in both A and B, then A and B are left relatively convex in $A *_C B$, by Example 1.18.

Example 1.22 Let A be a group, C a subgroup of A, and $x : C \to A$, $c \mapsto c^x$, be an injective homomorphism. The graph of groups $C \rightrightarrows A$, where the maps are the inclusion map and x, has as fundamental group

$$A *_C x := A * \langle x \mid \emptyset \rangle / \lhd \{x^{-1} \cdot c^{-1} \cdot x \cdot c^x \mid c \in C\} \rhd,$$

called the *HNN extension* with *vertex group* A, *edge group* C, and *edge map* x. We then view A and $\langle x \mid \emptyset \rangle$ as subgroups of $A *_C x$. In particular, $c^x = x^{-1}cx$.

If C and C^x are left relatively convex in A, then A is left relatively convex in $A *_C x$, by Theorem 1.17.

If $G = A *_C x$, then the Bass–Serre left G-tree T for $C \rightrightarrows A$ has vertex-set G/A and edge-set G/C, with $gA \xrightarrow{gC} gxA$.

1.4 Surface Groups and RAAGs

The following applies to all noncyclic compact surface groups.

Example 1.23 Let $G = \langle \{x\} \dot\cup \{y\} \dot\cup Z \mid x^{-1}y^{\epsilon}xyw \rangle$ with $\epsilon \in \{-1, 1\}$ and $w \in \langle Z \mid \emptyset \rangle$. By Example 1.20, both $\langle y \rangle$ and $\langle yw \rangle$ are left relatively convex in $\langle \{y\} \dot\cup Z \mid \emptyset \rangle$, which in turn is left relatively convex in the HNN extension G, by Example 1.22. Here the Bass–Serre left G-tree T has vertex-set $G/\langle \{y\} \cup Z \rangle$ and edge-set $G/\langle y \rangle$.

Notice that $\langle \{x\} \cup Z \rangle$ is not a left relatively convex subgroup in the group $\langle \{x\} \dot\cup \{y\} \dot\cup Z \mid (xy)^2 = x^2w^{-1} \rangle$.

Proposition 1.24 *(a) Every compact orientable surface group of genus $g \geq 1$ is locally n-indicable, for every $n \geq 1$.*

(b) Every compact non-orientable surface group of genus $g \geq 2$ is locally n-indicable, for $1 \leq n \leq g - 1$.

Proof (a) Let G be a compact orientable surface group of genus $g \geq 1$, $n \geq 1$, and let H be an m-generated subgroup of G that cannot be generated by fewer than m elements, for some $m \geq n$. If H is of infinite index in G, then H is free of rank m, and if it is of finite index, then it is a compact orientable surface group of genus $m/2 \geq g$. In both cases, H admits a homomorphism onto $\mathbb{Z}^m$, and therefore also onto $\mathbb{Z}^n$.

(b) Let G be a compact non-orientable surface group of genus $g \geq 2$, $n \geq 1$, and let H be an m-generated subgroup of G that cannot be generated by fewer than m elements, for some $m \geq n$. If H is of infinite index in G, then H is free of rank m, and if it is of finite index, then H is a compact non-orientable surface group of genus $m \geq g \geq n+1$. In the former case, H admits a homomorphism onto $\mathbb{Z}^m$, and therefore also onto $\mathbb{Z}^n$. In the latter case, H admits a homomorphism onto $\mathbb{Z}^{m-1}$ and, since in this case $m-1 \geq n$, H admits a homomorphism onto $\mathbb{Z}^n$. □

The following applies to all noncyclic surface groups except the Klein-bottle group.

Corollary 1.25
(a) Let G be a compact orientable surface group G of genus $g \geq 1$. Every maximal m-generated subgroup of G is left relatively convex in G.
(b) Let G be a compact non-orientable surface group G of genus $g \geq 3$. Every maximal m-generated subgroup of G, for $1 \leq m \leq g-2$ is left relatively convex in G.

Proof Follows from Proposition 1.24 and Corollary 1.8. □

Definition 1.26 Let X be a set, R be a subset of $[X, X]$ in $\langle X \mid \emptyset \rangle$, and $G = \langle X \mid R \rangle$. We say that G is a *right-angled Artin group*, or *RAAG* for short. For example, free groups and free abelian groups are RAAGs.

Let Y be a subset of X. The map $X \to G$ which acts as the identity map on Y and sends $X - Y$ to $\{1\}$ induces well-defined homomorphisms $G \to G$ and $G/\triangleleft X - Y \triangleright \to G$. Moreover, the natural composite $G/\triangleleft X - Y \triangleright \to G \to G/\triangleleft X - Y \triangleright$ is the identity map, since it acts as such on the generating set Y. Thus we may identify $G/\triangleleft X - Y \triangleright$ with its image $\langle Y \rangle$ in G. It follows that $\langle Y \rangle$ is a RAAG. We let $\pi_{\langle X \rangle \to \langle Y \rangle}$ denote the map $G \to G/\triangleleft X - Y \triangleright = \langle Y \rangle$.

For each $x \in X$, one has a splitting $G = A *_C x$, where $A = \langle X - \{x\} \rangle$, $C = \langle \{ y \in X - \{x\} \mid [x, y] \in R^{\pm 1} \} \rangle$, and $x : C \to A$, $c \mapsto c^x$, is the inclusion map. In essence, this was noted by Bergman [4].

It is not difficult to show that $\langle Y \rangle$ is left relatively convex in G; since (1.4.4) is a local condition, it suffices to verify this for X finite, and here it holds by induction on $|X|$ and Example 1.22. In particular, G is left orderable and, hence, torsion-free.

By [1, Corollary 1.6], RAAGs are nasmof and therefore locally 2-indicable. By Corollary 1.11, we have the following.

Corollary 1.27 *Let G be a subgroup a right-angled Artin group and κ a cardinal. Every maximal κ-generated abelian subgroup of G is left relatively convex in G.*

In particular, maximal abelian subgroups of G are left relatively convex in G.

Example 1.28 There exists an example, attributed by Minasyan [24] to Martin Bridson, of a subgroup G of finite index in the right-angled Artin group $F_2 \times F_2$ such that G/G' is not torsion-free. This implies that there exists $n \geq 2$ such that the right-angled Artin group $F_2 \times F_2$ is not locally n-indicable.

We do not know if RAAGs have the property that their maximal n-generated subgroups are left relatively convex.

1.5 Residually Torsion-free Nilpotent Groups and Left Relative Convexity

Corollary 1.8, combined with the next few observations, provides many examples of left relatively convex cyclic subgroups.

Proposition 1.29 *If G is a finitely generated, nilpotent group with torsion-free center, then G is 2-indicable.*

Proof Let G be any group (not necessarily nilpotent or with torsion free center), Z_1 be its center, and Z_2 be its second center, that is, Z_2/Z_1 is the center of G/Z_1.

For $g \in G$ and $a \in Z_2$, the commutator $[a, g]$ is in Z_1. From the identity $[ab, g] = [a, g]^b[b, g]$, we obtain, for $a, b \in Z_2$, $[ab, g] = [a, g][b, g]$. Therefore, for any element $g \in G$, $a \mapsto [a, g]$ is a homomorphism from Z_2 to Z_1, and $a \mapsto ([a, g])_{g \in G}$ is a homomorphism from Z_2 to $\prod_{g \in G} Z_1$ with kernel Z_1, which implies that Z_2/Z_1 embeds into a power of Z_1.

We now let G be a finitely generated, nilpotent group with torsion free center and we argue by induction on the nilpotency class c of G.

If $c = 0$, then G is trivial, and hence 2-indicable. Assume that $c \geq 1$. Since Z_2/Z_1 embeds into a power of Z_1, which is a torsion-free group, Z_2/Z_1 itself is a torsion-free group. Therefore G/Z_1 is a finitely generated, nilpotent group of class $c - 1$ with torsion-free center Z_2/Z_1. By the inductive hypothesis, G/Z_1 is 2-indicable. If G/Z_1 is noncyclic, then G/Z_1 maps onto $\mathbb{Z}^2$, and so does G; thus we may assume that G/Z_1 is cyclic. In that case, G/Z_1 is trivial, G is a free abelian group, and, hence, G is 2-indicable. □

Remark Note that, under the assumption that Z_1 is torsion free, the observation that Z_2/Z_1 embeds into some power of Z_1 yields that Z_2/Z_1, the center of G/Z_1, is itself torsion-free. Inductive arguments then quickly yield that each upper central series factor Z_{i+1}/Z_i, for $i \geq 0$, is torsion-free, each quotient Z_j/Z_i, for $j > i \geq 0$, is torsion free, and under the additional assumption that G is nilpotent, each quotient G/Z_i, for $i \geq 0$, is torsion-free; these are well-known results of Mal′cev [23] and we could use them to skip the first part in the proof of Proposition 1.29 and move directly to the inductive part of the proof.

Proposition 1.29 also follows from Mal′cev's result on quotients, together with Lemma 13 in [6], which states that every finitely generated, nilpotent group that is not virtually cyclic maps onto $\mathbb{Z}^2$ (the proof of this result relies on the fact that torsion-free, virtually abelian, nilpotent groups are abelian, which easily follows from the uniqueness of roots in torsion-free nilpotent groups; another result of Mal′cev from [23]).

With all these choices before us, we still opted for the proof of Proposition 1.29 provided above, because it is short and self-contained.

Proposition 1.30 *Every locally residually torsion-free nilpotent group is locally 2-indicable.*

Proof Let G be a locally residually torsion-free nilpotent group and H a finitely generated subgroup of G. Then H is a residually torsion-free nilpotent group. If H has a noncyclic, torsion-free, nilpotent quotient, then this quotient maps to $\mathbb{Z}^2$ by Proposition 1.29, and so does H. Otherwise, H is residually-$\mathbb{Z}$, which implies that it is abelian. Since H is finitely generated and torsion-free, it is free abelian, hence 2-indicable (in fact, H is cyclic in this case, since we already excluded the possibility of noncyclic quotients). □

Remark Note that if G is residually torsion-free nilpotent then it is also locally residually torsion-free nilpotent. In particular, for finitely generated groups there is no difference between being residually torsion-free nilpotent or being locally residually torsion-free nilpotent.

Example 1.31 If G is a

- residually free group [22],

- right-angled Artin group or a subgroup of a right-angled Artin group [14],
- 1-relator group with presentation

$$\langle\ X, a, b \mid [a, b] = w\ \rangle,$$

  where $a, b \notin X$ and w is a group word over X, including fundamental groups of all compact surfaces other than the sphere, the projective plane, and the Klein bottle [2, 17, 3],
- free group in any polynilpotent variety, including free solvable groups of any given class [18],
- pure braid group [16],
- 1-relator group with presentation

$$\langle x_1, \ldots, x_m, y_1, \ldots, y_n \mid u = v \rangle,$$

  where $v \in \langle y_1, \ldots, y_n \rangle$, $v \neq 1$, $u \in A = \langle x_1, \ldots, x_m \rangle$, $u \in \gamma_d(A)$ for some d such that u is not a proper power modulo $\gamma_{d+1}(A)$, where $\gamma_k(A)$ is the kth term of the lower central series of A [19],

then G is a residually torsion-free nilpotent group.

By Proposition 1.30, such a group G is locally 2-indicable and, by Corollary 1.8, each maximal cyclic subgroup of G is left relatively convex.

Acknowledgments

We thank Jack Button for his valuable input and, in particular, for pointing Example 1.28 to us. Corollary 1.15 is added at the suggestion of the referee.

Y.A. is supported by the Juan de la Cierva grant IJCI-2014-22425, and acknowledges partial support by Spain's Ministerio de Ciencia e Innovación through Project MTM2014-54896.

W.D. was partially supported by Spain's Ministerio de Ciencia e Innovación through Project MTM2014-53644.

Z.S. was partially supported by the National Science Foundation under Grant No. DMS-1105520.

References

[1] Antolín, Yago and Minasyan, Ashot. 2015. Tits alternatives for graph products. *J. Reine Angew. Math.*, **704**, 55–83.

[2] Baumslag, Gilbert. 1962. On generalised free products. *Math. Z.*, **78**, 423–438.

[3] Baumslag, Gilbert. 2010. Some reflections on proving groups residually torsion-free nilpotent. I. *Illinois J. Math.*, **54**(1), 315–325.

[4] Bergman, George M. 1976. *The global dimension of mixed coproduct/tensor-product algebras.* Preprint, 19 pages. Incorporated into [10] and [12].

[5] Bergman, George M. 1990. Ordering coproducts of groups and semigroups. *J. Algebra*, **133**(2), 313–339.

[6] Bridson, Martin R., Burillo, José, Elder, Murray and Šunić, Zoran. 2012. On groups whose geodesic growth is polynomial. *Internat. J. Algebra Comput.*, **22**(5), 1250048, 13.

[7] Burns, R. G. and Hale, V. W. D. 1972. A note on group rings of certain torsion-free groups. *Canad. Math. Bull.*, **15**, 441–445.

[8] Chiswell, I. M. 2011. Right orderability and graphs of groups. *J. Group Theory*, **14**(4), 589–601.

[9] Conrad, Paul. 1959. Right-ordered groups. *Michigan Math. J.*, **6**, 267–275.

[10] Dicks, Warren. 1981. An exact sequence for rings of polynomials in partly commuting indeterminates. *J. Pure Appl. Algebra*, **22**(3), 215–228.

[11] Dicks, Warren and Dunwoody, M. J. 1989. *Groups acting on graphs.* Cambridge Studies in Advanced Mathematics, vol. 17. Cambridge University Press, Cambridge. Errata at: `http://mat.uab.cat/~dicks/DDerr.html`.

[12] Dicks, Warren and Leary, I. J. 1994. Exact sequences for mixed coproduct/tensor-product ring constructions. *Publ. Mat.*, **38**(1), 89–126.

[13] Dicks, Warren and Šunić, Zoran. 2014. *Orders on trees and free products of left-ordered groups.* Preprint available at `http://arxiv.org/abs/1405.1676`.

[14] Droms, Carl. 1983. *Graph groups.* Ph.D. thesis, Syracuse University.

[15] Duncan, Andrew J. and Howie, James. 1991. The genus problem for one-relator products of locally indicable groups. *Math. Z.*, **208**(2), 225–237.

[16] Falk, Michael and Randell, Richard. 1988. Pure braid groups and products of free groups. Pages 217–228 of: *Braids (Santa Cruz, CA, 1986).* Contemp. Math., vol. 78. Amer. Math. Soc., Providence, RI.

[17] Frederick, Karen N. 1963. The Hopfian property for a class of fundamental groups. *Comm. Pure Appl. Math.*, **16**, 1–8.

[18] Gruenberg, K. W. 1957. Residual properties of infinite soluble groups. *Proc. London Math. Soc. (3)*, **7**, 29–62.

[19] Labute, John. 2015. *Residually torsion-free nilpotent one relator groups.* Preprint available at `http://arxiv.org/abs/1503.05167`.

[20] Linnell, Peter A., Rhemtulla, Akbar and Rolfsen, Dale P. O. 2009. Discretely ordered groups. *Algebra Number Theory*, **3**(7), 797–807.

[21] Louder, Larsen and Wilton, Henry. 2017. Stackings and the W-cycles conjecture. *Canad. Math. Bull.*, **60**(3), 604–612.

[22] Magnus, Wilhelm. 1935. Beziehungen zwischen Gruppen und Idealen in einem speziellen Ring. *Math. Ann.*, **111**(1), 259–280.

[23] Mal′cev, A. I. 1949. Nilpotent torsion-free groups. *Izvestiya Akad. Nauk. SSSR. Ser. Mat.*, **13**, 201–212.

[24] Minasyan, Ashot. 2012. Hereditary conjugacy separability of right-angled Artin groups and its applications. *Groups Geom. Dyn.*, **6**(2), 335–388.
[25] Ohnishi, Masao. 1952. Linear-order on a group. *Osaka Math. J.*, **4**, 17–18.
[26] Serre, Jean-Pierre. 1977. *Arbres, amalgames,* SL_2. Société Mathématique de France, Paris. Avec un sommaire anglais, Rédigé avec la collaboration de Hyman Bass, Astérisque, No. 46.
[27] Tychonoff, A. 1930. Über die topologische Erweiterung von Räumen. *Math. Ann.*, **102**(1), 544–561.

2

Groups with Context-free Co-word Problem and Embeddings into Thompson's Group V

Rose Berns-Zieve, Dana Fry, Johnny Gillings, Hannah Hoganson, and Heather Mathews

Abstract

Let G be a finitely generated group, and let Σ be a finite subset that generates G as a monoid. The *word problem of G with respect to Σ* consists of all words in the free monoid Σ^* that are equal to the identity in G. The *co-word problem of G with respect to Σ* is the complement in Σ^* of the word problem. We say that a group G is co$\mathcal{CF}$ if its co-word problem with respect to some (equivalently, any) finite generating set Σ is a context-free language.

We describe a generalized Thompson group $V_{(G,\theta)}$ for each finite group G and homomorphism θ: $G \to G$. Our group is constructed using the cloning systems introduced by Witzel and Zaremsky. We prove that $V_{(G,\theta)}$ is co$\mathcal{CF}$ for any homomorphism θ and finite group G by constructing a pushdown automaton and showing that the co-word problem of $V_{(G,\theta)}$ is the cyclic shift of the language accepted by our automaton.

A version of a conjecture due to Lehnert says that a group has context-free co-word problem exactly if it is a finitely generated subgroup of V. The groups $V_{(G,\theta)}$ where θ is not the identity homomorphism do not appear to have obvious embeddings into V, and may therefore be considered possible counterexamples to the conjecture.

Demonstrative subgroups of V, which were introduced by Bleak and Salazar-Diaz, can be used to construct embeddings of certain wreath products and amalgamated free products into V. We extend the class of known finitely generated demonstrative subgroups of V to include all virtually cyclic groups.

2.1 Introduction

Let G be a group and let $\Sigma \subseteq G$ be a finite set that generates G. The *word problem of G with respect to the free monoid Σ^** is the set of all words in Σ^* that are equivalent to the the identity in G. The *co-word problem of G with respect to Σ^** is the complement of the word problem. Both the word problem and the co-word problem of G are languages. The Chomsky Hierarchy [4] states that the set of regular languages is a subset of context-free languages, the set of context-free languages is a subset of context sensitive languages, and the set of context sensitive languages is a subset of recursive languages. We will focus in particular on context-free languages. A language is *context-free* if it is accepted by a pushdown automaton. If the co-word problem of G is a context-free language, then we say G is co$\mathcal{CF}$. This property does not depend on the choice of monoid generating set. The class of co$\mathcal{CF}$ groups was first studied by Holt, Rees, Röver, and Thomas [7]. They showed that the class is closed under taking finite direct products, taking restricted standard wreath products with virtually free top groups, and passing to finitely generated subgroups and finite index overgroups.

One group of particular interest is Thompson's group V, which is an infinite but finitely presented simple group. Lehnert and Schweitzer [11] demonstrate that Thompson's group V is co$\mathcal{CF}$. This group is of interest to us because of the conjecture, formulated by Lehnert [10] and revised by Bleak, Matucci, and Neunhöffer [2], that any group with context-free co-word problem embeds in V, i.e.,

Conjecture 2.1 [10, 2] Thompson's group V is a universal co$\mathcal{CF}$ group.

In this chapter we prove two classes of result, one related to embeddings into V, and the other offering a potential counterexample to Conjecture 2.1.

Bleak and Salazar-Diaz [1] define the class of demonstrative subgroups of V and use this class to produce embeddings of free products and wreath products into V. They also show that the class of groups that embed into V is closed under taking finite index overgroups. Their proof of the latter fact appeals to results of Kaloujnine and Krasner [9]. Here, we use induced actions to give a direct proof. Our argument shows, moreover, that if the original embedding is demonstrative, then so is the embedding of the finite index overgroup.

A theorem of [1] says that $\mathbb{Z}$ is a demonstrative subgroup of V. The results sketched above prove that all virtually cyclic groups are demonstrative, and it appears that these are the only known finitely generated demonstrative subgroups. If V is a universal co$\mathcal{CF}$ group, then it should be possible, by the results of Holt, Rees, Röver, and Thomas [7], to find an embedding of $G \wr F_2$ into V, where G is co$\mathcal{CF}$ and F_2 is the free group on two generators. The easiest way to find such an embedding would be to show that F_2 has a demonstrative embedding into V. We are thus led to ask:

Question *Does there exist a demonstrative embedding of F_2 into V?*

Our class of potential counterexamples to Conjecture 2.1 comes from the cloning systems of Witzel and Zaremsky [14]. We look at a specific group $V_{(G,\theta)}$ that arises from their family of groups equipped with a cloning system. We define a surjective homomorphism Φ from $V_{(G,\theta)} \to V$, which implies that $V_{(G,\theta)}$ acts on the Cantor set. However, by our construction, $V_{(G,\theta)}$ seems to have no obvious faithful actions on the Cantor set when θ is not the identity homomorphism.

In our main result, we prove that $V_{(G,\theta)}$ is co$\mathcal{CF}$ for all pairs of θ and finite G. We begin by detailing a construction of a pushdown automaton and we show that the co-word problem is equivalent to the cyclic shift of the language accepted by the automaton, therefore proving that the co-word problem is context-free.

We briefly outline the chapter. Section 2.2 provides the necessary background for the reader to understand the concepts discussed in the two following sections. In Section 2.3, we give our proofs and examples of all ideas related to demonstrative subgroups. Finally, in Section 2.4 we prove our main result.

2.2 Background

2.2.1 Pushdown Automata

Definition 2.2 Let Σ be a finite set, called an *alphabet*. We call elements of the alphabet *symbols*. The *free monoid* on Σ, denoted Σ^*, is the set of all finite strings of symbols from Σ. This includes the empty string, which we denote ϵ. The operation is concatenation. An element of the free monoid is a *word*. A subset of the free monoid is a *language*.

Example 2.3 Let $\Sigma = \{0, 1\}$. The free monoid Σ^* contains all finite concatenations of 0 and 1 in any order. An example word is 01101.

Definition 2.4 Let G be a group. A *finite monoid generating set* is a finite alphabet Σ with a surjective monoid homomorphism $\Phi : \Sigma^* \to G$. The *word problem* of a group G (with respect to Σ), denoted $WP_\Sigma(G)$, is the kernel of Φ. The complement of the word problem is the *co-word problem*, denoted $CoWP_\Sigma(G)$.

Definition 2.5 Let Σ, Γ be alphabets and let $\#$ be an element of Γ. A *pushdown automaton* [4] with stack alphabet Γ and input alphabet Σ is defined as a directed graph with a finite set of vertices V, a finite set of *transitions* (directed edges) δ, an initial state $v_0 \in V$, and a set of *terminal states* $T \subseteq V$.

A transition is labeled by an ordered triplet $(w_1, w_2, w_3) \in (\Sigma \cup \{\epsilon\}) \times \Gamma^* \times \Gamma^*$. When *following* a transition, the pushdown automaton (PDA) reads and deletes w_1 from its input tape, reads and deletes w_2 from its memory stack (shortened to stack for the duration of this chapter), and writes w_3 on its stack. If w_1, w_2, or w_3 equals ϵ, the automaton does not execute the action associated with that coordinate. We only consider a *generalized* PDA, which (as described above) can add and delete multiple letters on its stack at a time. A PDA accepts languages either by terminal state, or by empty stack. This must be specified upon creation of the automaton. See Definitions 2.7 and 2.8.

Definition 2.6 ([6], Definition 2.6) Let P be a pushdown automaton. We describe a class of directed paths issuing from the basepoint v_0 in P, called the *valid paths*, by induction on length. We simultaneously (inductively) define the *stack values* of valid paths. The path of length 0 starting at the initial vertex $v_0 \in P$ is valid; its stack value is $\# \in \Gamma^*$. Let $t_1 \dots t_n (n \geq 0)$ be a valid path in P, where t_1 is the transition crossed first. Let t_{n+1} be a transition whose initial vertex is the terminal vertex of t_n; we suppose that the label of t_{n+1} is (s, w_1, w_2). The path

Figure 2.1 The stack of a PDA. As an element is read from the top, the next element appears as the new top if nothing is written to the stack.

$t_1 \ldots t_n t_{n+1}$ is also valid, provided that the stack value of $t_1 \ldots t_n$ has w_1 as a prefix; that is, if the stack value of $t_1 \ldots t_n$ has the form $w_1 w' \in \Gamma^*$ for some $w' \in \Gamma^*$. We say that the edge t_{n+1} is a *valid transition*. The stack value of $t_1 \ldots t_n t_{n+1}$ is then $w_2 w'$. We let *val(p)* denote the stack value of a valid path p.

The *label* of a valid path $t_1 \ldots t_n$ is $s_n \ldots s_1$, where s_i is the first coordinate of the label for t_i (an element of Σ, or the empty string). The label of a valid path p will be denoted $\ell(p)$.

Definition 2.7 Let p be a valid path of a pushdown automaton P. We say that p is a *successful path* when

1 $val(p) = \epsilon$ if P accepts by empty stack, or
2 The terminal vertex of p is in T if P accepts by terminal state.

Definition 2.8 Let P be a PDA. A word $w \in \Sigma^*$ is *accepted* by P if w is the label $\ell(p)$ of a successful path p. The language *accepted by* P, denoted $\mathcal{L}_P$, is the collection of all words accepted by the automaton; i.e.,

$$\mathcal{L}_P = \{w \in \Sigma^* \mid w = \ell(p) \text{ for some successful path } p\}.$$

Definition 2.9 A subset of the free monoid Σ^* is called a *(non-deterministic) context-free language* if it is $\mathcal{L}_P$ for some pushdown automaton P.

Let P be a PDA. We operate P in the following way.

1 A word $\widehat{w}$ is placed on the input tape and read symbol by symbol. By our convention, P reads the input tape from right to left.
2 Next, P follows valid transitions non-deterministically (i.e., by choosing them) until $\widehat{w} = \ell(p)$ for some successful path p. Throughout this process, the leftmost symbol on the stack is considered to be in the top position.
3 If some successful path p exists, P accepts $\widehat{w}$.

2.2.2 Thompson's Group V

Definition 2.10 Let $X = \{0, 1\}$ and consider X^*. As in [1], we define an infinite rooted tree, $\mathcal{T}_2$, as follows.

The set of nodes for $\mathcal{T}_2$ is X^*. For $u, v \in X^*$, there exists an edge from u to v if $ux = v$ for some $x \in X$.

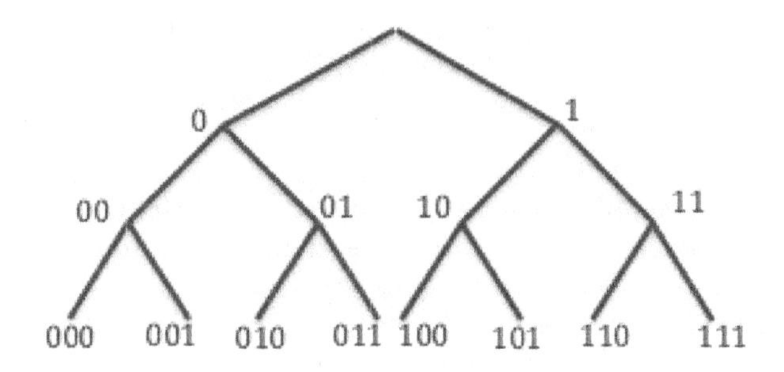

Figure 2.2 The top portion of the infinite binary tree $\mathcal{T}_2$ with some of its nodes labeled.

Definition 2.11 An *infinite path* $\mathcal{T}_2$ is an infinite string of 0s and 1s. $Ends(\mathcal{T}_2)$ is the collection of all such infinite paths.

We say that $u \in X^*$ is a *prefix* of $\omega \in Ends(\mathcal{T}_2)$ if there is $\hat{\omega} \in Ends(\mathcal{T}_2)$ such that $\omega = u\hat{\omega}$. For $u \in X^*$, we let $u* = \{\omega \in Ends(\mathcal{T}_2) \mid u \text{ is a prefix of } \omega\}$.

Define $d : Ends(\mathcal{T}_2) \times Ends(\mathcal{T}_2) \to \mathbb{R}$ by $d(\zeta_1, \zeta_2) = e^{-l}$, where $\zeta_1, \zeta_2 \in Ends(\mathcal{T}_2)$ and l is the length of the longest prefix shared by ζ_1 and ζ_2. The function d is a metric on $Ends(\mathcal{T}_2)$.

For $w \in Ends(\mathcal{T}_2)$, let $B_r(w) = \{\zeta \in Ends(\mathcal{T}_2) \mid d(\zeta, w) \leq r\}$. Then $B_r(w)$ is the *metric ball around w with radius r*. It can be shown that each metric ball in $Ends(\mathcal{T}_2)$ takes the form $u*$, for some $u \in X^*$.

We note that $Ends(\mathcal{T}_2)$ is a Cantor set. Thompson's group V acts as self-homeomorphisms on this Cantor set, and each element of V can be represented by a binary tree pair. Furthermore, the leaves of the trees can be represented in binary code where a branch to the left is denoted by "0" and a branch to the right by "1".

The group V is generated by the maps A, B, C, and π_0 [5]. We define the generators of V by the prefix changes they represent, which are equivalent to the tree diagrams in Figure 2.3. For instance, if $\omega \in Ends(\mathcal{T}_2)$ has the form $\omega = 0\hat{\omega}$, for some $\hat{\omega} \in Ends(\mathcal{T}_2)$, then $A(\omega) = 00\hat{\omega}$.

A	B	C	π_0
$0* \mapsto 00*$	$0* \mapsto 0*$	$0* \mapsto 11*$	$0* \mapsto 10*$
$10* \mapsto 01*$	$10* \mapsto 100*$	$10* \mapsto 0*$	$10* \mapsto 0*$
$11* \mapsto 1*$	$110* \mapsto 101*$	$11* \mapsto 10*$	$11* \mapsto 11*$
	$111* \mapsto 11*$		

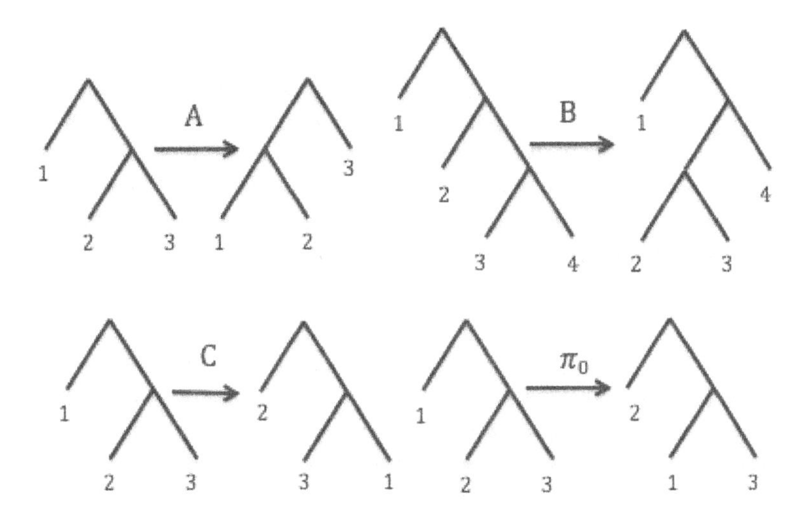

Figure 2.3 Elements A, B, C, and π_0 of V represented as tree pairs.

2.2.3 Generalized Thompson Groups $V_{(G,\theta)}$

Definition 2.12 [14] The *forest monoid*, $\mathcal{F}$, consists of all sequences of ordered, rooted, binary trees $(T_i)_{i\in\mathbb{N}}$, where all but finitely many trees are trivial (i.e., consist only of the root). For two elements $E_1, E_2 \in \mathcal{F}$, their product, E_1E_2, is obtained by attaching the ith leaf of E_1 with the ith root of E_2.

Given a finite group, G, define

$$H = S_\infty \ltimes_\phi (\oplus_{i=1}^{\infty} G),$$

where $\phi : S_\infty \to Aut(\oplus_{i=1}^{\infty} G)$ is defined by

$$\phi(\sigma)(g_1, ..., g_k, ...) = (g_{\sigma^{-1}(1)}, ...g_{\sigma^{-1}(k)}, ...)$$

for $\sigma \in S_\infty$.

Multiplication of the pair of group elements $(\sigma_1, (g_1, ..., g_k, ...))$ and $(\sigma_2, (g'_1, ..., g'_k, ...)) \in H$ is defined as follows:

$$\begin{aligned}(\sigma_1, (g_1, ..., g_k, ...))(\sigma_2,(g'_1, ..., g'_k, ...)) \\ &= (\sigma_1\sigma_2, \phi(\sigma_2)(g_1, ..., g_k, ...)(g'_1, ..., g'_k)) \\ &= (\sigma_1\sigma_2, (g_{\sigma_2^{-1}(1)}g'_1, ..., g_{\sigma_2^{-1}(k)}g'_k, ...)).\end{aligned}$$

Our group $V_{(G,\theta)}$ arises from the cloning system construction in [14]. The idea behind cloning systems is to give a general method for producing hybrids between Thompson's group F (see [5] for an introduction) and other families of groups. One example of such a hybrid is the braided Thompson group BV introduced by Brin [3]. Here BV is a hybrid between F and the infinite-strand braid group $B_\infty = \bigcup_{n=1}^{\infty} B_n$. The group V itself is another example; it is a hybrid of F and $S_\infty = \bigcup_{n=1}^{\infty} S_n$. In the formal theory, a *cloning system* consists of a group H, a homomorphism $\rho : H \to S_\infty$, and a collection of cloning maps $\{\kappa_k \mid k \in \mathbb{N}\}$. The cloning maps formally describe how to multiply pairs of the form (f, h), where $h \in H$ and $f \in \mathcal{F}$. A cloning system that satisfies conditions $CS1, CS2, CS3$ of Proposition 2.7 of [14] defines a Brin–Zappa–Szep product, $\mathcal{F} \bowtie H$, which is something like a semidirect product. Witzel and Zaremsky [14] show that $\mathcal{F} \bowtie H$ is a cancellative monoid with least common right multiples. The generalized Thompson group determined by the cloning system is the group of right fractions of $\mathcal{F} \bowtie H$. We refer the reader to [14] for a full discussion.

Elements of $V_{(G,\theta)}$ are ordered pairs of elements from a subgroup of the BZS product defined by the following cloning system:

$$H = S_\infty \ltimes_\phi (\oplus_{i=1}^{\infty} G),$$

$$\rho(\sigma, (g_1, ..., g_k)) = \sigma,$$

where κ_k acts on the right by:

$$(\sigma, (g_1, ..., g_k, g_{k+1}, ...))\kappa_k = (\sigma\zeta, (g_1, ..., g_k, \theta(g_k), g_{k+1}, ...)),$$

where ζ is the cloning map for the symmetric group defined in Example 2.9 of [14], and θ is an arbitrary homomorphism from $G \to G$.

We can think of elements of $V_{(G,\theta)}$ as equivalence classes of tree pairs much as in Thompson's group V. The difference is that tree pairs in

$V_{(G,\theta)}$ have group elements from G attached to their leaves. A tree pair (a, b) can be modified within its equivalence class by adding canceling carets or canceling group elements.

Canceling carets are added to corresponding leaves of the domain and range trees just as they would be in Thompson's group V, unless there is a group element g on the leaf, in which case we put a g on the left branch of the new caret and $\theta(g)$ on the right branch. Canceling group elements are added to corresponding leaves of the domain and range trees and are combined, using the group operation, with any group element already on the leaves. We can multiply two elements $(a, b), (c, d) \in V_{(G,\theta)}$, by choosing equivalent tree pairs (a', b') and (c', d') where $b' = c'$. Then $(a, b)(c, d) = (a', b')(c', d') = (a', d')$. (Note that, in these tree pairs, the second coordinate corresponds to the domain, and the first to the range.)

Note that any element $(a, b) \in V_{(G,\theta)}$ can be expressed with no group elements in the domain tree by adding canceling group elements.

If $\{g_1, ..., g_n\}$ is a generating set for G, then

$$\{A, B, C, \pi_0\} \cup \{g_{ja}, g_{jb}, g_{jc}, g_{jd}, g_{je} \mid 1 \leq j \leq n\}$$

is a generating set for $V_{(G,\theta)}$, where $g_{ja}, g_{jb}, g_{jc}, g_{jd}, g_{je}$ are defined as in Figure 2.6.

Remark If $\theta = id_G$, the identity homomorphism, then $V_{(G,\theta)}$ embeds in V. We will sketch a proof of this fact. The first step is to construct an embedding of $V_{(G,\theta)}$ into the group of homeomorphisms of $G \times \mathcal{C}$, where $\mathcal{C}$ denotes the set $Ends(\mathcal{T}_2)$.

Let (a, b) be a tree pair representing an element of $V_{(G,\theta)}$; we can arrange that the domain tree b has no nontrivial elements of G attached to its leaves. We define a function $\alpha_{(a,b)} : \mathcal{C} \to G$ as follows. If $a_0a_1\ldots$ lies under the ith leaf of the domain tree b, then $\alpha_{(a,b)}(a_0a_1\ldots)$ is defined to be the group element $g \in G$ that is attached to the ith leaf of the range tree a. For instance, if (a, b) is the tree pair representing g_{je} from Figure

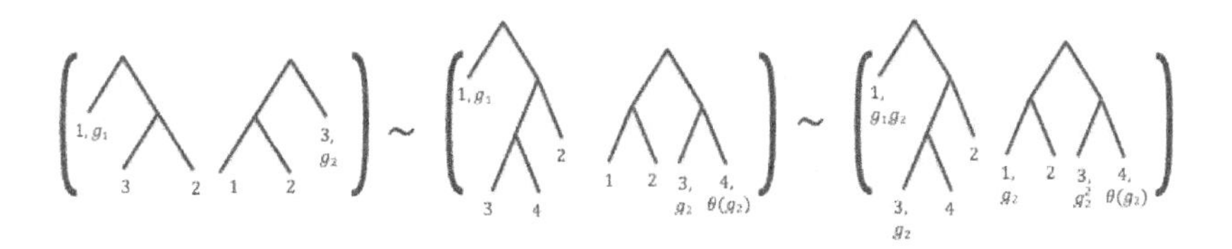

Figure 2.4 Three equivalent tree pairs, the second is obtained by adding a canceling caret and the third by canceling group elements.

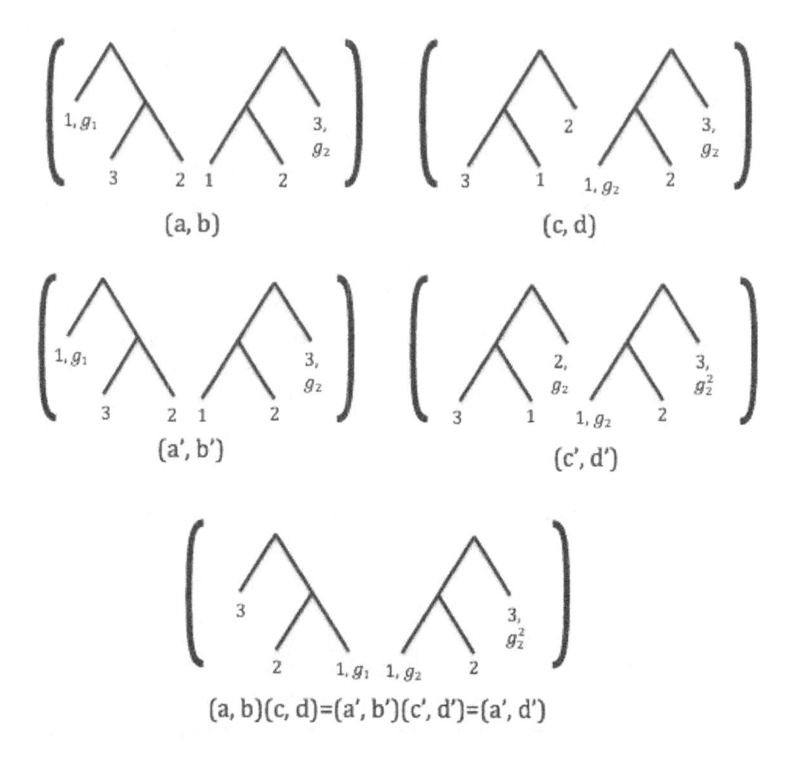

Figure 2.5 Multiplication of two group elements (a,b) and (c,d).

2.6, then $\alpha_{(a,b)}(111\ldots) = g_j$, but $\alpha_{(a,b)}(110\ldots) = 1_G$, since $110\ldots$ lies under the third leaf of the domain tree, and the third leaf of the range tree is labeled by the trivial element.

Let $\pi_{(a,b)}$ be the element of V that is obtained by removing any elements of G from the leaves of the range tree a. For each tree pair (a,b), we define a homeomorphism $h_{(a,b)} : G \times \mathcal{C} \to G \times \mathcal{C}$ as follows:

$$h_{(a,b)}(g, \zeta) = (\alpha_{(a,b)}(\zeta)g, \pi_{(a,b)}(\zeta)).$$

It is possible to show that $h_{(a,b)}$ does not change if pairs of canceling carets are added or removed to the tree pair (a,b). (This uses the fact that θ is the identity homomorphism of G.) It follows that the assignment $(a,b) \to h_{(a,b)}$ determines a function $\phi : V_{(G,\theta)} \to Homeo(G \times \mathcal{C})$; a check shows that ϕ is an injective homomorphism. A proof of injectivity goes like this: if $\pi_{(a,b)} \neq 1$, then $h_{(a,b)}$ is non-trivial, since the second coordinate will be acted upon non-trivially. If $\pi_{(a,b)} = 1$ but (a,b) does

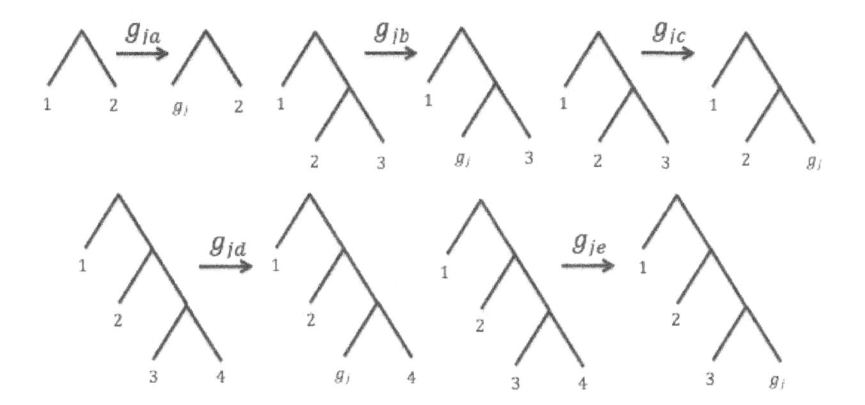

Figure 2.6 Additional generators of $V_{(G,\theta)}$. Note that, in this figure, the domain trees are on the left.

not represent the trivial element in $V_{(G,\theta)}$, then $\alpha_{(a,b)}(\zeta) \neq 1_G$ for some $\zeta \in \mathcal{C}$, so

$$h_{(a,b)}(1_G, \zeta) = (\alpha_{(a,b)}(\zeta), \zeta) \neq (1_G, \zeta).$$

Next, we identify $G \times \mathcal{C}$ with $\mathcal{C}$ as follows. Choose a partition $\mathcal{P}$ of $\mathcal{C}$ into $|G| = n$ disjoint open balls: $\mathcal{P} = \{w_1*, \ldots, w_n*\}$. Define $I : G \times \mathcal{C} \to \mathcal{C}$ by sending (g_k, ζ) to $w_k\zeta$. (Note that the latter is the concatenation of a finite string with an infinite sequence.) This map is easily seen to be a homeomorphism.

Finally, we define the embedding $\psi : V_{(G,\theta)} \to V$ by the rule $\psi((a, b)) = I \circ h_{(a,b)} \circ I^{-1}$.

2.3 Demonstrative Groups

Definition 2.13 [1] Suppose a group G acts by homeomorphisms on a topological space X. For a group $H \leq G$, the action of H in G is *demonstrative* if and only if there exists an open set $U \subset X$ such that for all $h_1, h_2 \in G$, $h_1 U \cap h_2 U \neq \emptyset$ if and only if $h_1 = h_2$. The set U is called a *demonstration set.*

For this discussion, let $G = V$ and $X = Ends(\mathcal{T}_2)$. In this case, if U (as above) is a metric ball (i.e. $U = w*$ for some $w* \subseteq Ends(\mathcal{T}_2)$), then we refer to U as a *demonstration node.*

Definition 2.14 Let $H \leq G$ and let H act on a topological space S. We define $G \times_H S$ to be $\{(g, s) : g \in G, s \in S\}/\sim$, where $(gh, s) \sim (g, h \cdot s)$ for all $h \in H$.

The *induced action* of G on $G \times_H S$ is $* : G \times (G \times_H S) \to G \times_H S$ defined by $g_1 * (g_2, s) = (g_1 g_2, s)$.

Bleak and Salazar-Díaz [1] proved that the class of subgroups of V is closed under passage to finite-index overgroups; i.e., given an embedding $H \to V$, there is an embedding $G \to V$ if H has finite index in G. Here, we give a different proof of this fact. Our method also shows that the class of demonstrative subgroups of V is closed under passage to finite-index overgroups.

Theorem 2.15 *Let H be a subgroup of Thompson's group V. If $H \leq G$ where $[G : H] = m$, for some $m \in \mathbb{N}$, then G embeds in V. Moreover, if H embeds as a demonstrative subgroup in V, then G embeds as a demonstrative subgroup of V.*

Proof Assume $H \leq G$. Choose a left transversal $T = \{t_1, t_2, ..., t_m\}$ for H in G with $t_1 = 1$. We can induce an action of G on $G \times_H Ends(\mathcal{T}_2)$ by:

$$g \cdot (t_i, x) = (gt_i, x), \text{ for } x \in Ends(\mathcal{T}_2).$$

We know we can write gt_i as $t_j h$ for some unique $t_j \in T$ and $h \in H$. So,

$$(gt_i, x) = (t_j h, x) = (t_j, h \cdot x).$$

Now, we can embed $G \times_H Ends(\mathcal{T}_2) \hookrightarrow Ends(\mathcal{T}_2)$ by choosing a set $W = \{w_1, \ldots, w_m\}$ such that $\{w_1*, w_2*, \ldots, w_m*\}$ is a fixed partition of $Ends(\mathcal{T}_2)$, and defining an injective function $\phi : T \to W$ by $\phi(t_i) = w_i$. Now define $\Phi : G \times_H Ends(\mathcal{T}_2) \to Ends(\mathcal{T}_2)$ by $\Phi((t_i, x)) = \phi(t_i)x = w_i x$ and $\Phi((g \cdot t_i, x) = g \cdot \Phi(t_i, x)$.

It can be easily checked that

$$g \cdot (w_i x) = g \cdot \Phi(t_i, x) = \Phi(gt_i, x) = \Phi(t_j, h \cdot x) = w_j h(x)$$

is a group action of G on $Ends(\mathcal{T}_2)$. Additionally, because elements of H act as elements of V and $[G : H] < \infty$, so do elements of G. Therefore, G embeds in V.

Now, assume H has a demonstrative embedding in V with demonstration node $a_1 a_2 ... a_n *$ for $a_i \in \{0, 1\}$.

We will show that $w_i a_1 \ldots a_n *$ is a demonstration node for G. We compute the action of each of $g, g' \in G$ on $w_i a_1 \dots a_n *$:

$$g \cdot w_i a_1 \dots a_n * = w_j h(a_1 \dots a_n *),$$
$$g' \cdot w_i a_1 \dots a_n * = w_k h'(a_1 \dots a_n *).$$

Since $w_1 *, \dots, w_m *$ partition $Ends(\mathcal{T}_2)$, if $w_j * \cap w_k * \neq \emptyset$ then we have $j = k$. Since H is a demonstrative subgroup of G with demonstration node $a_1 a_2 \dots a_n *$, $h(a_1 \dots a_n *) \cap h'(a_1 \dots a_n *) \neq \emptyset$ if and only if $h = h'$. Thus, $w_j h(a_1 \dots a_n *) \cap w_k h'(a_1 \dots a_n *) \neq \emptyset$ if and only if $j = k$ and $h = h'$, in other words, if and only if $g = g'$. Thus, G is demonstrative in V with demonstration node $w_i a_1 \dots a_n *$ for any $i \in \{1, ..., m\}$. □

2.4 Main Result

The proof of our main theorem uses a result (Proposition 2.17) that is implicit in both [6] and [11].

Definition 2.16 [6, Definition 3.1] Let $S = \{A, B, C, \pi_0\}$ be the generating set for V from Figure 2.3. Let $T = \{B_1, \dots, B_k\}$ be a finite partition of $Ends(\mathcal{T}_2)$ into open balls. We say that T is a *test partition* if, for any word $w = s_1 \dots s_n$ ($s_i \in S \cup S^{-1}$) such that $w \neq 1$ in V, there is some cyclic permutation $s' = s_i s_{i+1} \dots s_n s_1 \dots s_{i-1}$ and a ball $B_j \in T$ such that $s'(B_j) \neq B_j$.

Proposition 2.17 *([6]; [11]) The partition*

$$T = \{a_0 a_1 a_2 * \mid a_i \in \{0, 1\} \textit{ for } (i = 0, 1, 2)\}$$

is a test partition.

Proof First, we notice that, for each generator $s \in \{A, B, C, \pi_0\} \cup \{A, B, C, \pi_0\}^{-1}$, each ball $a_0 a_1 a_2 *$ is contained in some region for s. (Here a *region* of s is a ball in $Ends(\mathcal{T}_2)$ that corresponds to a leaf in the domain tree of s; see Definition 2.19 in [6].) It follows that we may take T as the partition $\mathcal{P}_{big}$ in Definition 3.3 of [6]. Since the group Γ in Definition 3.5 of [6] is always trivial, we may also take T as the partition $\mathcal{P}_{small}$ in Definition 3.5 of [6]. The desired conclusion now follows from Proposition 3.6 of [6]. □

Lemma 2.18 *Let* $\Sigma = \{A, B, C, \overline{A}, \overline{B}, \pi_0, g_{ij}\}$, *where* $i \in \{1, \dots, n\}$, *and $j \in \{a, b, c, d, e\}$. Define $\Phi : \Sigma^* \to V$ by the surjective homomorphism*

$$\begin{aligned}\Phi : A &\mapsto A\\ \bar{A} &\mapsto A^{-1}\\ B &\mapsto B\\ \bar{B} &\mapsto B^{-1}\\ C &\mapsto C\\ \pi_0 &\mapsto \pi_0\\ g_{ij} &\mapsto 1_V.\end{aligned}$$

If $w = b_1 \cdots b_m \in \Sigma^*$ satisfies $\Phi(w) \neq 1$, then there is a cyclic permutation $b_j \cdots b_m b_1 \cdots b_{j-1}$ and some $\widehat{B} \in \{a_1a_2a_3* : a_i \in 0, 1\}$ that satisfy $b_j \cdots b_m b_1 \cdots b_{j-1}(\widehat{B}) \cap \widehat{B} \neq \emptyset$. (Here the action of a word $w \in \Sigma^*$ on the ball $\widehat{B}$ is determined by the rule $w(\widehat{B}) = \Phi(w)(\widehat{B})$.)

Proof Consider $w = b_1 \cdots b_m \in \Sigma^*$. If $\Phi(w) \neq 1$, then we have $\Phi(w) \in CoWP(V)$. We know $\{a_1a_2a_3* : a_i \in 0, 1\}$ is a test partition for V by Proposition 2.17. So, there is some cyclic permutation $\Phi(w)'$ of $\Phi(w)$ and some $\widehat{B} \in \{a_1a_2a_3* : a_i \in \{0, 1\}\}$ such that $\Phi(w)'(\widehat{B}) \cap \widehat{B} \neq \emptyset$. Since Φ takes all the generators g_{ij} to 1, Φ will preserve the shape of any tree pair. Thus, if $\Phi(w)'$ is such that $\Phi(w)'(\widehat{B}) \cap \widehat{B} \neq \emptyset$, then $w'(\widehat{B}) \cap \widehat{B} \neq \emptyset$ where w' is some cyclic shift $b_j \cdots b_m b_1 \cdots b_{j-1}$ of w. □

Definition 2.19 Let $\mathcal{L}$ be a language. The *cyclic shift* of $\mathcal{L}$, denoted $\mathcal{L}^\circ$, is

$$\mathcal{L}^\circ = \{w_2w_1 \in \Sigma^* \mid w_1w_2 \in \mathcal{L}, w_1, w_2 \in \Sigma^*\}.$$

A *cyclic permutation* w' of a word $w = w_1w_2$ is $w' = w_2w_1$. Note that the class of context-free languages is closed under cyclic shifts [12].

Theorem 2.20 *The group $V_{(G,\theta)}$ is co$\mathcal{CF}$.*

Proof Let $G = \{g_1, \ldots, g_n\}$. We design an automaton P to accept by empty stack, with stack alphabet $\Gamma = \{0, 1, g \mid g \in G\backslash\{1_G\}\}$ and input alphabet $\Sigma = \{A, B, C, \overline{A}, \overline{B}, \pi_0, g_{ij}\}$, where $i \in \{1, \ldots, n\}$, $j \in \{a, b, c, d, e\}$, and the g_{ij} are as defined in Figure 2.6.

We define

$$\mathcal{L}_{B_k} = \{w \in \Sigma^* \mid w(B_k) \cap B_l \neq \emptyset \text{ for some } l \neq k\},$$

where the B_k and B_l are balls from the partition $T = \{a_0a_1a_2* \mid a_i \in \{0, 1\}\}$, and the action of a word $w \in \Sigma^*$ on the ball B_k is determined by the rule $w(B_k) = \Phi(w)(B_k)$, as in Lemma 2.18. We let $\mathcal{L}_G$ be the set of words w in Σ^* such that there is a tree pair representative for $w \in V_{(G,\theta)}$ with no group elements written on the leaves of the domain tree, and at least one non-trivial $g \in G$ written on a leaf of the range tree.

We design P such that $\mathcal{L}_P = (\bigcup_{B_k \in T} \mathcal{L}_{B_k}) \cup \mathcal{L}_G$. Figure 2.7 outlines a portion of the automaton. Note that unlabeled arrows represent $(\epsilon, \epsilon, \epsilon)$ transitions. From the initial loading phase, there are in fact eight different arrows $(\epsilon, \epsilon, B_k)$, one for each of the $B_k \in \{000, 001, \ldots, 111\}$. These lead to eight separate reading and accept phases, each as pictured in the figure. These reading and accept phases are identical, with one exception: the labels on the arrows leading to the test partition accept state vary. For instance, in the accept phase corresponding to 000, the arrow labeled $(\epsilon, B_l, \epsilon)$ corresponds to seven different arrows, one for each $B_l = a_1a_2a_3$, where $a_i \in \{0, 1\}$ and not all of a_1, a_2, a_3 are 0.

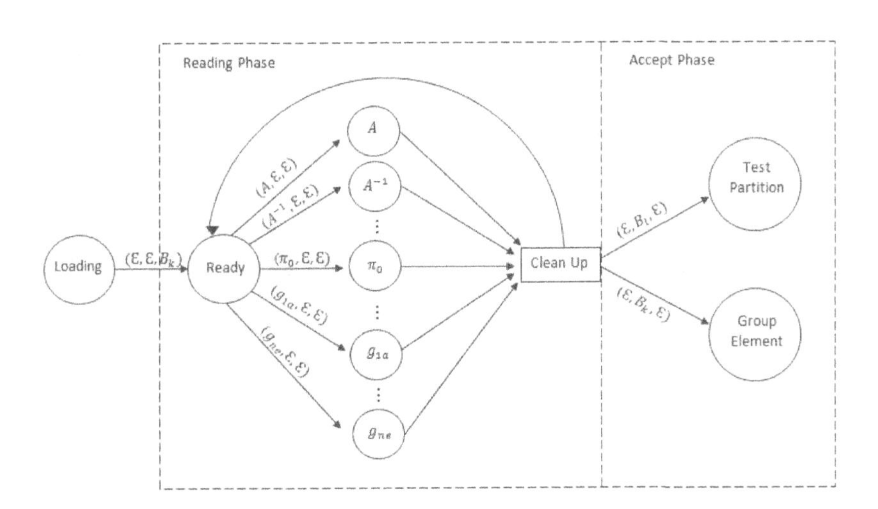

Figure 2.7 Sample reading and accept phases of the automaton for $V_{(G,\theta)}$.

To start off, P enters a non-deterministic loading phase. This consists of a single state S with transitions labeled $(\epsilon, \epsilon, 0)$ and $(\epsilon, \epsilon, 1)$, both leading back to S. Here a finite string of 1s and 0s is entered non-deterministically onto the stack.

We leave the loading phase by taking a transition $(\epsilon, \epsilon, B_k)$ where

$$B_k \in \{000, 001, 010, 100, 011, 110, 101, 111\},$$

i.e. B_k is one of the eight metric balls in the test partition.

Next, P enters the reading phase which has a single state for each generator, where the first element on the input tape is read and that generator is applied to the appropriate prefix at the top of the stack. For example, the reading phase for A would read and delete a 0, and then add 00; the reading phase for g_{2b} would read and delete 10, and then add $10g_2$.

After the reading phase, P enters the clean-up state, which consists of the pushing and combining of stack elements. This phase allows P to "clean-up" the stack so that there are at least three 0s and 1s for the next element on the input tape to successfully act on the stack. First, P "pushes" elements within the first three spots to the fourth spot on the stack. For example, as shown in Figure 2.8, one set of transitions will read and delete $0g0$ or $0g1$ from the stack and then add $00g$ or $01(\theta(g))$, respectively, to the stack for all $g \in G\backslash\{1_G\}$. Similar transitions can be followed if the group element is preceded by the prefix 0, 1, 01, 10, 00, or 11.

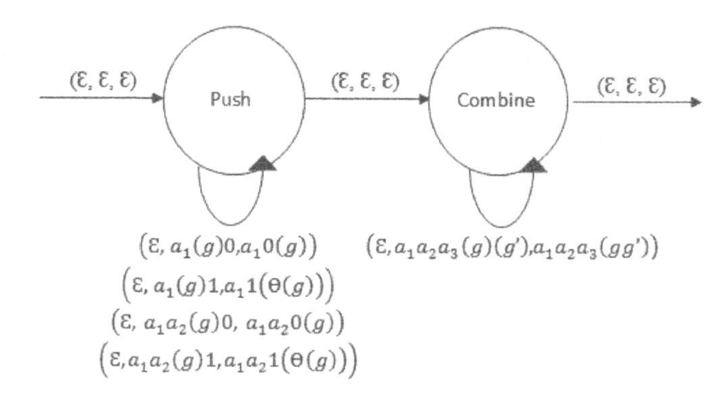

Figure 2.8 The clean-up phase. Let $a_1, a_2, a_3 \in \{0, 1\}$ and $g, g' \in G\backslash\{I_G\}$.

Next, P enters the combining state where group elements are rewritten as a single element of the group (in accordance with the group operation). For example, one collection of edges is able to read and delete $010(g)(g')$ and add $010(gg')$ for all $g, g' \in G\backslash\{1_G\}$. Note that if $gg' = 1_G$, then the path reads and deletes $010(g)(g')$, and writes 010. Similar paths exist when combining any two group elements preceded by any three-digit prefix.

After exiting the clean-up phase, P reads the next element on the input tape and repeats the process of the reading phase. When the input tape is empty and P has gone through the reading and clean-up phases, P then moves onto one of two accept states.

If the word from the input tape took one metric ball B_k in the test partition to some other metric ball B_l, then the three-letter prefix describing B_l is now showing on the stack, so we can follow a path labeled $(\epsilon, B_l, \epsilon)$ to the *test partition accept state.* (Here we recall that the single arrow labeled $(\epsilon, B_l, \epsilon)$ in Figure 2.7 is actually seven different arrows, one for each l such that $B_l \neq B_k$.) At this point we "unload" all of the 0s and 1s and group elements off the stack with paths (ϵ, x, ϵ) for $x \in \{0, 1, g \mid g \in G \backslash \{1_G\}\}$. Once every stack element has been deleted, P takes the path $(\epsilon, \#, \epsilon)$ which deletes the start symbol and accepts the word. So the language accepted by the eight test partition accept states is $(\bigcup_{B_k \in T} \mathcal{L}_{B_k})$.

If the word from the input tape does not displace metric ball B_k, then we enter the *group element accept state.* Here, we delete every 0 and 1 on the stack until P arrives at a group element. The group element is then"pushed" further down the stack, and the 0 or 1 it pushes past is deleted. For example, one path is $(\epsilon, g0, g)$ for $g \in G \backslash \{1_G\}$. If the group element is followed on the stack by a second group element, then they are "combined" in a manner mimicking the previously described combining portion of the clean-up phase. This is repeated until there are no 0s or 1s left on the stack. At this point, if there is still a group element remaining on the stack followed by the start symbol, then they are both deleted and thus the word is accepted. However, if there is no group element on the stack then the start symbol cannot be deleted so the word is not accepted. Assuming that the address of an appropriate metric ball was written on the stack in the loading phase, there will be a group element remaining, and so the language accepted by the eight group element accept states is $\mathcal{L}_G$.

Thus, $\mathcal{L}_P = (\bigcup_{B_k \in T} \mathcal{L}_{B_k}) \cup \mathcal{L}_G$.

We claim that $CoWP(V_{(G,\theta)}) = (\mathcal{L}_P)^\circ$.

Let $w \in \mathcal{L}_P$, so that $w \in \mathcal{L}_G$ or $w \in \mathcal{L}_{B_k}$, for some k. If $w \in \mathcal{L}_G$, then it follows directly that $w \in CoWP(V_{(G,\theta)})$. If $w \in \mathcal{L}_{B_k}$ for some k, then $w(B_k) \cap B_l \neq \emptyset$ for $l \neq k$ $(B_k, B_l \in T)$, so $w \in CoWP(V_{(G,\theta)})$. Therefore, $\mathcal{L}_P \subseteq CoWP(V_{(G,\theta)})$. The $CoWP$ of a group is closed under cyclic shift, and thus $(\mathcal{L}_P)^\circ \subseteq CoWP(V_{(G,\theta)})$.

Let $w \in CoWP(V_{(G,\theta)})$. We will use the surjective homomorphism Φ from Lemma 2.18. If $w \notin Ker(\Phi)$, then $\Phi(w) \neq 1_V$. By Lemma

2.18, there is some cyclic permutation w' and some $B_k \in T$ such that $w'(B_k) \cap B_k \neq \emptyset$, i.e. $w' \in \mathcal{L}_{B_k}$. Therefore, $w \in (\mathcal{L}_{B_k})^\circ \subseteq (\mathcal{L}_P)^\circ$. If $w \in Ker(\Phi)\backslash\{1_{V_{(G,\theta)}}\}$, then $\Phi(w) = 1_V$, which implies that w (as a reduced tree pair) does not change any prefixes; it simply adds group elements. Therefore, $w \in \mathcal{L}_G \subseteq (\mathcal{L}_P)^\circ$.

We now have that $CoWP(V_{(G,\theta)}) = (\mathcal{L}_P)^\circ$. A language is context-free if and only if its cyclic shift is also context-free. Since $\mathcal{L}_P$ is context-free, $CoWP(V_{(G,\theta)})$ is context-free, and $V_{(G,\theta)}$ is co$\mathcal{CF}$.

□

Remark We attempted a similar method of proof with the group generated by Thompson's group V and the Grigorchuk group G, Rövers's group [13] $R = \langle V, G\rangle$. Like $V_{(G,\theta)}$, elements of R can be thought of as Thompson's group V elements with Grigorchuk group elements, g attached to the leaves. However, the Grigorchuk group elements continue to act on the tree whereas the group elements in $V_{(G,\theta)}$ just sit at the end of the branches. This became a problem because it is impossible to complete the calculation of the action of g on any finite test string loaded onto an automaton. We also ran into problems because the test partitions argument used in Theorem 6.1. and for Finite Similarity Structure Groups [8] fails.

Acknowledgments

We would like to first thank our research advisor, Dr. Dan Farley, for aiding us in our research endeavors and providing us with the knowledge needed for us to be successful. We thank the Summer Undergraduate Mathematical Science Research Institute (SUMSRI) at Miami University in Oxford, OH, for giving us the opportunity to participate in undergraduate research in mathematics, and we thank the faculty and staff for making this program possible. In addition, we would like to express our gratitude to Miami University for providing funding, housing, and a place to do research. Finally, we thank the National Science Foundation for funding SUMSRI and making our research possible.

References

[1] Bleak, Collin and Salazar-Díaz, Olga. 2013. Free products in R. Thompson's group V. *Trans. Amer. Math. Soc.*, **365**(11), 5967–5997.

[2] Bleak, Collin, Matucci, Francesco and Neunhöffer, Max. 2016. Embeddings into Thompson's group V and $coC\mathcal{F}$ groups. *J. London Math. Soc. (2)*, **94**(2), 583–597.

[3] Brin, Matthew G. 2007. The algebra of strand splitting. I. A braided version of Thompson's group V. *J. Group Theory*, **10**(6), 757–788.

[4] Brookshear, J. Glenn. 1989. *Formal Languages, Automata, and Complexity*. Theory of Computation. The Benjamin/Cummings Publishing Company.

[5] Cannon, J. W., Floyd, W. J. and Parry, W. R. 1996. Introductory notes on Richard Thompson's groups. *Enseign. Math. (2)*, **42**(3-4), 215–256.

[6] Farley, Daniel. 2014. *Local similarity groups with context-free co-word problem*. This volume. Preprint available at `http://arxiv.org/abs/1406.4590`.

[7] Holt, Derek F., Rees, Sarah, Röver, Claas E. and Thomas, Richard M. 2005. Groups with context-free co-word problem. *J. London Math. Soc. (2)*, **71**(3), 643–657.

[8] Hughes, Bruce. 2009. Local similarities and the Haagerup property. *Groups Geom. Dyn.*, **3**(2), 299–315. With an appendix by Daniel S. Farley.

[9] Kaloujnine, Léo and Krasner, Marc. 1950. Produit complet des groupes de permutations et problème d'éxtension de groupes. I. *Acta Sci. Math. Szeged*, **13**, 208–230.

[10] Lehnert, J. 2008. *Gruppen von quasi-Automorphismen*. Ph.D. thesis, Goethe Universität, Frankfurt.

[11] Lehnert, J. and Schweitzer, P. 2007. The co-word problem for the Higman-Thompson group is context-free. *Bull. Lond. Math. Soc.*, **39**(2), 235–241.

[12] Maslov, A. N. 1973. The cyclic shift of languages. *Problemy Peredači Informacii*, **9**(4), 81–87.

[13] Röver, Claas E. 1999. Constructing finitely presented simple groups that contain Grigorchuk groups. *J. Algebra*, **220**(1), 284–313.

[14] Witzel, Stefan and Zaremsky, Matt. 2018. Thompson groups for systems of groups, and their finiteness properties. *Groups Geom. Dyn.*, **12**(1), 289–358.

3

Limit Sets for Modules over Groups Acting on a $CAT(0)$ Space

Robert Bieri and Ross Geoghegan

Abstract

This is a summary, written by the first-named author, of my joint work with Ross Geoghegan over the past years. Most of the material is available in detail in [4], and I will occasionally refer to specific detail in that paper. Other parts of our joint work – results mostly concerned with extending concepts and results from [4] to higher dimensions – will also be mentioned but are still in preparation. I am very pleased for this opportunity to thank my friend Ross for his excellent, enjoyable and fruitful collaboration over more than 15 years. Obviously, I cannot dedicate this joint summary to him ... but do it with what I found in the process of writing:

I dedicate to Ross the conjectures in Section 3.5

on the occasion of his 70th birthday

(while at the same time keeping the responsibility for their accuracy).

3.1 Horospherical Limit Sets of G-modules

G-modules as well as actions of the group G by isometries on $CAT(0)$ metric spaces are major tools in group theory. Here we consider the situation when a group G comes together with two such G-objects: a pair (M, A) consisting of a *proper* G-$CAT(0)$-space and a *finitely generated* G-module A. In this generality we relate the G-module A to the G-space M by means of a G-map

$$L : A \to \{\text{subsets of } \partial M\}.$$

where ∂M is the boundary of M at infinity which carries the G-action induced by the action on M.

To define L we choose (1) a free presentation $\epsilon : F \twoheadrightarrow A$, where F is the free G-module over the finite basis X, and (2) a base point $b \in M$. For each $c \in F$ we consider the *support*, $\mathrm{supp}(c) \subseteq G$: the (finite) set of elements of G occurring in the unique expansion of c over GX. Then putting $h(c) := \mathrm{supp}(c)b \subseteq M$ defines a G-map $h : F \to fM$, from F into the G-space fM of all finite subsets of M. We call h a *control map* and define the *horospherical limit points* of $a \in A$ to be the points in the set $L(a)$, defined to be

$$\{e \in \partial M \mid \text{every horoball at } e \text{ contains } h(c) \text{ for some } c \in \epsilon^{-1}(a)\}.$$

We show that $L(a)$ is independent of the choice of $\epsilon : F \twoheadrightarrow A$ and the base point b. By the *horospherical limit set of* A we mean the set of all points $e \in \partial M$ which are limit points of all elements of A: i.e. $\Sigma(M; A) := \bigcap_{a \in A} L(a)$.

The main point of our work is generality; and our foremost interest in generality is the range of $CAT(0)$ spaces M: from the Euclidean to the hyperbolic. Thus we are primarily interested in accessible G-spaces which – intriguingly often – come naturally together with important specific groups. Examples we have in mind are the natural action of an abstract group G on $G_{\mathrm{ab}} \otimes \mathbb{R}$ by left translations, or groups which come together with a linear or fractional-linear transformation over ($\mathbb{R}$, $\mathbb{C}$ or the p-adic numbers) – and hence have well-known natural actions on hyperbolic spaces, symmetric spaces, trees or buildings.

3.2 Two General Results on the G-pairs (M, A)

Theorem 3.1 $\Sigma(M; A) = \partial M$ *if and only if the G-action on M is cocompact and A is supported over a bounded subset B of M (i.e. for each $a \in A$ there is some $c \in F$ with $\epsilon(c) = a$ and $h(c) \subseteq B$).*

Corollary 3.2 *If the G-action on M is cocompact with discrete orbits then $\Sigma(M; A) = \partial M$ if and only if A is finitely generated as a module over a point stabilizer.*

Theorem 3.3 *If $\Sigma(M; A) = \partial M$ holds for a given isometric action $\rho : G \to \mathrm{Isom}(M)$, then it also holds in a neighborhood of ρ in the space* $\mathrm{Hom}(G, \mathrm{Isom}(M))$.

Corollary 3.4 (a) *The set of cocompact actions of G on M by isometries is open.* (b) *Let $R(G, M) \subseteq \mathrm{Hom}(G, \mathrm{Isom}(M))$ be the subspace of all actions such that $\rho(G)$ is cocompact with discrete orbits and finite stabilizers. For every finitely generated G-module A the set*

$$\{\rho \mid A \text{ is finitely generated over } \ker(\rho)\}$$

is open in $R(G, M)$.

3.3 A Unifying Concept

Our proof of Corollary 3.4 requires all the subtle details of the theory, and thus the result testifies, in some sense, to the value of $\Sigma(M; A)$. Personally, I am more motivated by the intriguing fact that $\Sigma(M; A)$ and its close relatives show up in a number of highly interesting mathematical contexts.

3.1 If G acts with discrete orbits on M and trivially on $A \neq 0$ then our limit set $\Sigma(M; A) \subseteq \partial M$ is the usual horospherical limit set of the orbits. A variant of $\Sigma(M; A)$ which we denote $\Lambda(M; A)$ coincides with the classical limit set of G in that case.

3.2 If $\mathbf{G}$ is the Lie group $\mathrm{SL}(n, \mathbb{R})$ acting on its symmetric space $M = \mathrm{SL}(n, \mathbb{R})/\mathrm{SO}(n)$ then the building at infinity $B(\mathbf{G})$ comes with a natural surjection $\pi : B(\mathbf{G}) \twoheadrightarrow \partial M$, compatible with the Tits metric, and in that case the horospherical limit set of $\mathbf{G}$ (with respect to the trivial $\mathbf{G}$-module $\mathbb{Z}$) is the whole of ∂M. When we restrict the action to the arithmetic subgroup $G = \mathrm{SL}(n, \mathbb{Z})$, the horospherical limit set is much smaller: $\Sigma(M; \mathbb{Z}) \subseteq \partial M$ is now the complement of the image $\pi(B(G)) \subseteq \partial M$ of the rational building. This was a conjecture of Hanno Rehn (who studied the higher homotopical version of our limit sets [3] in his thesis [10] in the case when $G = \mathrm{SL}(n, \mathbb{Z}[\frac{1}{m}])$). It has recently been proved for more general arithmetic groups by G. Avramidi and D. Witte-Morris [1].

3.3 In the case when M is Euclidean and G acts by a discrete translation group $\Sigma(M; A)$ is the 0-dimensional part of the Bieri–Neumann–Strebel–Renz invariant, a group theoretic tool for questions related to homological finiteness properties of infinite groups. (In fact, the present work grew out of the aim to extend the leading ideas of the BNSR theory to the $CAT(0)$ case.)

(a) The special case where M is Euclidean and G is abelian is intimately connected with tropical geometry. Here, the complement of our horospherical limit set appears as the radial projection of the integral tropical variety associated with the annihilator ideal of the G-module A.

(b) The 1-dimensional part of the BNS invariant, denoted $\Sigma^1(G) = \Sigma^1(G;\mathbb{Z})$, is not, in general, a horospherical limit set in the above sense; but in a subsequent paper it will appear as a special case of a higher dimensional version. Here we recall that when G is the fundamental group of a 3-manifold X, $\Sigma^1(G)$ can be interpreted as the set of interior points of the top dimensional faces of the (polyhedral) unit-ball associated with the Thurston norm in $H^1(X;\mathbb{Z})$.

Why I think this is worth our effort. To consider limit sets of a $CAT(0)$ G-space associated with G-modules is new. It plays a crucial role in the Euclidean case and its applications. The potential for a general theory relating G-modules to the geometry of $CAT(0)$ G-spaces is evident in view of 3.2 and the dominant role of modules and the symmetric space for linear groups.

Moreover, my experience of treating $\Sigma(M;A)$ respectfully as an object of interest in its own right is encouraging.

(a) By computing its complement in [5], tropical varieties (along with their polyhedrality and substantial parts of the "Fundamental Theorem of Tropical Geometry" were detected long before their official date of birth.

(b) The conjecture in Hanno Rehn's thesis opened the view on the intriguing (and now established) fact that the complement of $\Sigma(M;\mathbb{Z})$ – a G-invariant subspace of ∂M which is defined in full generality – carries, in the special case when G is a nice arithmetic group, the full information of the rational building at infinity.

3.4 G-dynamical Limit Points and G-finitary Homomorphisms

In order to prove Theorems 3.1 and 3.3 we had to consider subsets of $\Sigma(M;A)$ with a dynamical flavor. This requires measuring the quality of the convergence of a sequence of finite subsets of M towards a boundary point $e \in \partial M$ in terms of the Busemann function[1] $\beta_e : M \to \mathbb{R}$.

Let $\epsilon : F \twoheadrightarrow A$ be the free presentation of Section 3.1. We say that $e \in \partial M$ is a *G-dynamical limit point of* the pair (M, A) if there is a G-endomorphism $\varphi : F \to F$ which induces the identity of A and has the property that there is a number $\delta > 0$ with $\min\beta_e(h(\varphi(c))) \geq \min\beta_e(h(c)) + \delta$ for all $c \in F$. Note that $h(\varphi^i(c))$ exhibit e as a horospherical limit point of $\epsilon(c)$, hence the *G-dynamical limit set*

$$^{\circ\circ}\Sigma(M; A) := \{e \in \partial M \mid e \text{ is a } G\text{-dynamical limit point of } (M, A)\}$$

is a subset of $\Sigma(M; A)$. The concept is not new: in the case of a Euclidean discrete action it was a crucial lemma ([8], [9] and [7]) that $^{\circ\circ}\Sigma(M; A) = \Sigma(M; A)$. But in the general $CAT(0)$ case $^{\circ\circ}\Sigma(M; A)$ is often dramatically smaller than $\Sigma(M; A)$ as we can see from the following theorem:

Theorem 3.5 *If $e \in {}^{\circ\circ}\Sigma(M; A)$ then the closure (in the cone topology) of the orbit Ge, $\mathrm{cl}_{\partial M}(Ge)$, lies in a Tits-metric ball with radius $r < \frac{\pi}{2}$. Moreover, the center of the unique minimal ball with this property is fixed by G.*

The joy of this nice fixed point theorem is somewhat spoiled when we face the fact that the most interesting group actions (like any non-elementary Fuchsian group on the hyperbolic plane) do not have fixed points in ∂M. Thus $^{\circ\circ}\Sigma$ will be empty in these cases and hence cannot be a useful tool for actions on hyperbolic spaces. This suggests that considering dynamical limit sets only when $\varphi : F \to F$ is a G-endomorphism is too restrictive. Instead we had to find a class of additive endomorphisms $\varphi : F \to F$ more flexible than G-endomorphisms but still sharing some of their coarse features.

Definition 3.6 (*G-finitary homomorphisms*). An additive homomorphism $\varphi : A \to B$ between G-modules is *G-finitary* if there is a G-map $\Phi : A \to fB$ of the G-set underlying A into the G-set fB of all finite subsets of B with the property that $\varphi(a) \in \Phi(a)$ for every $a \in A$. We say that φ is a *selection* from the *G-volley* Φ.

Definition 3.7 We say that e is a *(finitary) dynamical limit point* of the pair (M, A) if there is a G-finitary endomorphism $\varphi : F \to F$, which induces the identity of A and has the property that there is a number $\delta > 0$ with $\min\beta_e(h(\varphi(c))) \geq \min\beta_e(h(c)) + \delta$ for all $c \in F$.

[1] Our convention is that $\beta_e(b) = 0$ for a fixed base point $b \in M$, and $\beta_e(e) = +\infty$.

Note that G-endomorphisms are G-finitary, hence every G-dynamical limit point is dynamical. The set

$$^\circ\Sigma(M;A) := \{e \in \partial M \mid e \text{ is a dynamical limit point of } (M,A)\}$$

is the main technical tool of the chapter. Its precise relationship to the horospherical limit set is given in the following Theorem.

Theorem 3.8 *$^\circ\Sigma(M;A)$ is a G-invariant subset of $\Sigma(M;A)$; it consists of all $e \in \Sigma(M,A)$ with the property that $\mathrm{cl}(Ge)$, the cone-topology-closure of the G-orbit of e, is contained in $\Sigma(M;A)$. In particular, the set $^\circ\Sigma(M;A)$ contains every closed G-invariant subset of $\Sigma(M;A)$, and hence $^\circ\Sigma(M;A) = \partial M$ if and only if $\Sigma(M;A) = \partial M$.*

3.5 $^\circ\Sigma(M;A)$ as an Object of Interest in its Own Right

I believe that the dynamical invariant $^\circ\Sigma(M;A)$ qualifies as a member of this league for the following reasons.

(1) It combines cone topology and the Tits metric topology of ∂M in an interesting way. On the one hand *$^\circ\Sigma(M;A)$ is a Tits-open subset of ∂M*; on the other hand *if $e \in {}^\circ\Sigma(M;A)$ then $^\circ\Sigma(M;A)$ contains not only the orbit Ge but also the cone-topology closure of that orbit*[2] $\mathrm{cl}_{\partial M}(Ge)$.
(2) In the G-finitary category of G-modules[3] the Fundamental Theorem of Homological Algebra holds true: every G-finitary homomorphism between two modules A and B can be lifted to a G-finitary chain map between the projective resolutions of A and B, and any two lifts are homotopic by a G-finitary homotopy. That is precisely what we need to extend the definition of $^\circ\Sigma(M;A)$ to higher-dimensional invariants $^\circ\Sigma^n(M;A)$ when A admits a free resolution with finite n-skeleton, and to prove our openness results for those. This will appear in a subsequent paper.
(3) For each $e \in \partial M$ we consider the set $\widehat{\mathbb{Z}G}^e$ of all formal sums $\Sigma_{g\in G} n_g g$ with integer coefficients n_g, and the property that for each horoball H at e the set $\{g \in G \mid n_g \neq 0 \text{ and } gb \notin H\}$ is finite. We observe that $\widehat{\mathbb{Z}G}^e$ is a right G-module which contains the group ring as a

[2] See Theorem 3.8.
[3] The category whose objects are G-modules and whose morphisms are G-finitary additive maps.

submodule, and we call it the *Novikov module* at e. Then we have the following.

Theorem 3.9 *$e \in {}^{\circ}\Sigma^{n}(M; A)$ if and only if $\mathrm{Tor}_k^{\mathbb{Z}G}(\widehat{\mathbb{Z}G^{e'}}, A) = 0$ for all $0 \leq k \leq n$ and all e' contained in the closure of the orbit Ge in ∂M.*

This is useful since it opens the possibility of relating the various ${}^{\circ}\Sigma^{n}(M; A)$ via the long exact Tor sequences. It indicates that ${}^{\circ}\Sigma(M; A)$ is perhaps better behaved that $\Sigma(M; A)$ with respect to the module argument.

(4) **Polyhedrality questions.** As mentioned in 3.3(a), the case when G is finitely generated and abelian acting on Euclidean $G \otimes \mathbb{R}$ intersects with tropical geometry. Here ${}^{\circ\circ}\Sigma$, ${}^{\circ}\Sigma$ and Σ coincide, and their complement in the sphere $\partial(G \otimes \mathbb{R})$ is closely related to (a $\mathbb{Z}$-version of) Bergman's logarithmic limit set [2]. An interpretation of the definitions in terms of matrices shows that the dynamical invariants ${}^{\circ\circ}\Sigma(M; A)$ and ${}^{\circ}\Sigma(M; A)$ constitute this relationship. Now, Bergman conjectured, and a theorem at the roots of tropical geometry proved in [5] establishes, that $\Sigma(G \otimes \mathbb{R}; A)$ is a rational polyhedral subset of the sphere (the proof is outlined in the appendix of [4]).

Even though there is no reasonable notion of "polyhedrality" for subsets at infinity of general $CAT(0)$-spaces I claim that the following is a straightforward generalization of the Bergman conjecture: Given a G-volley $\Phi : F \to fF$ let $\Sigma(\Phi)$ denote the set of all $e \in \partial M$ which are in ${}^{\circ}\Sigma(M; A)$ witnessed by selections from this one volley Φ.

Conjecture 3.10 There is a G-volley $\Phi : F \to fF$ with $\Sigma(\Phi) = {}^{\circ}\Sigma(M; A)$.

Proof of the claim If G is finitely generated abelian, Theorem B of [6] yields a finite set of elements $\Lambda \subseteq \mathbb{Z}G$ with the property that right multiplication of each $\lambda \in \Lambda$ into the free module F lifts the identity of A, and the G-map $\Phi : F \to fF$, $\Phi(c) = c\Lambda$ is the required volley. □

In the case when G is a finitely generated abelian group and $M = G \otimes \mathbb{R}$ is Euclidean, hence ${}^{\circ\circ}\Sigma(M; A) = {}^{\circ}\Sigma(M; A)$ is the original Σ-invariant in the sphere at infinity of M, the conjecture readily implies the more plausible but still wide open

Conjecture 3.11 The BNSR-Invariant $\Sigma^0(G; A)$ is always rational-polyhedral.

(5) The fact that $^\circ\Sigma(M; A)$ and $^{\circ\circ}\Sigma(M; A)$ are independent of the particular free presentation of A shows that the above-mentioned matrix interpretation of these invariants is a condition on stable matrices. Whether experience with stable matrix theory – possibly under the assumption that the Whitehead group is trivial – can be of help to exploit this flexibility remains to be seen.

References

[1] Avramidi, Grigori and Morris, Dave Witte. 2014. Horospherical limit points of finite-volume locally symmetric spaces. *New York J. Math.*, **20**, 353–366.

[2] Bergman, George M. 1971. The logarithmic limit-set of an algebraic variety. *Trans. Amer. Math. Soc.*, **157**, 459–469.

[3] Bieri, Robert and Geoghegan, Ross. 2003. Connectivity properties of group actions on non-positively curved spaces. *Mem. Amer. Math. Soc.*, **161**(765), xiv+83.

[4] Bieri, Robert and Geoghegan, Ross. 2016. Limit sets for modules over groups on CAT(0) spaces – from the Euclidean to the hyperbolic. *Proc. London Math. Soc. (3)*, **112**(6), 1059–1102.

[5] Bieri, Robert and Groves, J. R. J. 1984. The geometry of the set of characters induced by valuations. *J. Reine Angew. Math.*, **347**, 168–195.

[6] Bieri, Robert and Harlander, Jens. 2001. A remark on the polyhedrality theorem for the Σ-invariants of modules over abelian groups. *Math. Proc. Cambridge Philos. Soc.*, **131**(1), 39–43.

[7] Bieri, Robert and Renz, Burkhardt. 1988. Valuations on free resolutions and higher geometric invariants of groups. *Comment. Math. Helv.*, **63**(3), 464–497.

[8] Bieri, Robert and Strebel, Ralph. 1980. Valuations and finitely presented metabelian groups. *Proc. London Math. Soc. (3)*, **41**(3), 439–464.

[9] Bieri, Robert, Neumann, Walter D. and Strebel, Ralph. 1987. A geometric invariant of discrete groups. *Invent. Math.*, **90**(3), 451–477.

[10] Rehn, Wolf Hanno. 2007. *Kontrollierter Zusammenhang über symmetrischen Räumen.* Ph.D. thesis, Johann Wolfgang Goethe-Universität in Frankfurt am Main.

4

Ideal Structure of the C^*-algebra of R. Thompson's group T

Collin Bleak and Kate Juschenko

Abstract

We explore the ideal structure of the reduced C^*-algebra of R. Thompson's group T. We show that even though T has trace, one cannot use the Kesten Condition to verify that the reduced C^*-algebra of T is simple. At the time of the initial writing of this chapter, there was no example of a group for which it was known that the Kesten Condition would fail to prove simplicity, even though the group has trace. Motivated by this first result, we describe a class of groups where even if the group has trace, one cannot apply the Kesten Condition to verify the simplicity of those groups' reduced C^*-algebras. We also offer an apparently weaker condition to test for the simplicity of a group's reduced C^*-algebra, and we show this new test is still insufficient to show that the reduced C^*-algebra of T is simple. Separately, we find a controlled version of a Ping-Pong Lemma which allows one to find non-abelian free subgroups in groups of homeomorphisms of the circle generated by elements with rational rotation number. We use our Ping-Pong Lemma to find a simple converse to a theorem of Uffe Haagerup and Kristian Knudsen Olesen which shows that R. Thompson's group F will be non-amenable if and only if there exists a finite set $H \subset T$ which can be decomposed as disjoint union of sets H_1 and H_2 with $\sum_{g\in H_1} \pi(g) = \sum_{h\in H_2} \pi(h)$ and such that the closed ideal generated by $\sum_{g\in H_1} \lambda(g) - \sum_{h\in H_2} \lambda(h)$ coincides with $C^*_\lambda(T)$.

4.1 Introduction

This chapter is predominately about dynamical properties of groups of homeomorphisms, and on testing when groups of homeomorphisms with unique trace can be shown to have reduced C^*-algebras. Much progress has been made in this area since this chapter was first drafted, but we believe our initial results remain of interest (see the remarkable papers [16, 3, 19, 17]).

Our initial motivation for beginning the work described here arose out of interest in the amenability question for R. Thompson's group F, and the beautiful paper [12] of Haagerup and Olesen (see also [20, Section 4.2]).

4.1.1 On Simplicity

Since the foundational paper [21] of Powers, there has been a long-standing interest in whether there could exist a group G with unique trace, for which its reduced C^*-algebra $C^*_\lambda(G)$ is not simple. The paper [19] answers this affirmatively, and in particular the papers [16, 3, 19, 17] fully describe the implication relationships between C^*-simplicity, being tracial, and having an amenable radical.

In this chapter, we will focus on a method from [21] in which Powers gives the following test for the simplicity of the algebra $C^*_\lambda(G)$ over a group G.

Theorem 4.1 (Powers) *If for all non-empty $H \subset G$ with $|H| < \infty$, $e \notin H$ and for all positive integers n there is a set $\{c_1, c_2, \ldots, c_n\} \subset G$ so that*

$$\lim_{n\to\infty} \frac{1}{n}||\Sigma_{i=1}^{n}\lambda(c_i h {c_i}^{-1})|| = 0, \; \forall h \in H,$$

then $C^*_\lambda(G)$ is simple.

Let G be a group generated by a finite set S with $S = S^{-1}$, then $\frac{1}{|S|}||\Sigma_{h\in S}\lambda(h)||$ is equal to the spectral radius of the simple random walk on the Cayley graph of G with respect to S, denoted by $\rho(G, S)$. The spectral radius of the simple random walk has been computed for many groups. Kesten in [18] shows that if $S = \{g_1, \ldots, g_n\}$ is *a free set*, i.e., $g_1, \ldots, g_n$ are standard generators of the free group of rank n, then the spectral radius is

$$\rho(G, S) = \frac{\sqrt{2n-1}}{n}.$$

Thus the following condition implies the hypothesis of Theorem 4.1.

Condition 4.2 (Kesten Test) *For all finite subsets $H \subset G$ with $e \notin H$ and for all positive integers n there is a set $\{c_1, c_2, \ldots, c_n\} \subset G$ so that*

$$\langle h^{c_1}, h^{c_2}, \ldots, h^{c_n} \rangle$$

is a free subgroup of G of rank n for all $h \in H$.

Previous to the results of [16, 3, 19, 17], the Kesten Test has been a main test for detecting the simplicity of the reduced C^*-algebra of a group, and indeed, one of the results of [17] is that a group has a simple reduced C^*-algebra if and only if it satisfies the hypotheses of Powers' Theorem. We show here that the Kesten Test for detecting when the hypotheses of Powers' Theorem hold is actually inadequate for many groups of permutations (e.g., groups of homeomorphisms of spaces).

To explain the comments above, we need to set some notation. If g is a bijection from a set X to itself, denote by $Supp(g) := \{x \in X \mid g \cdot x \neq x\}$ and $Fix(g) := X \backslash Supp(g)$, the support and the set of points fixed by g, respectively.

The following lemma shows the inadequacy of Condition 4.2 for groups of permutations of a set X.

Lemma 4.3 *Let X be a set, and G the group of bijections from X to X. Suppose h_1, $h_2 \in G \backslash \{1\}$ so that $Supp(h_1) \cap Supp(h_2) = \emptyset$. If c_1, $c_2 \in G$ so that $Supp(h_1^{c_1}) \cup Supp(h_1^{c_2}) = X$ then*

$$Supp(h_2^{c_1}) \cap Supp(h_2^{c_2}) = \emptyset.$$

The reader can prove the above lemma with little effort, but for completeness we include a proof in the next section. In particular, the following theorem is a corollary to Lemma 4.3.

Theorem 4.4 *Let G be a group of permutations of some set X so that any finite collection of elements of G can only represent a free basis if the union of their supports is the entire set X. Suppose that $H \subset G \backslash \{1_G\}$ admits elements h_1 and h_2 so that $Supp(h_1) \cap Supp(h_2) = \emptyset$. Then for $n \geq 2$ there is no set of elements $\{c_1, c_2 \ldots, c_n\}$ so that $\langle h^{c_1}, h^{c_2}, \ldots, h^{c_n} \rangle$ is a free group on n generators for all $h \in H$.*

Now by Theorem 4.4, Condition 4.2 fails to apply to Thompson's group T and many other groups with unique trace. In particular, groups of homeomorphisms of spaces which contain any pair of non-trivial elements with disjoint supports, and where a subset of the elements can only be a free basis if the union of their supports is the entire set X, can never satisfy Condition 4.2. Thus, we offer a new test (Condition 4.5) which is a natural apparent weakening of Condition 4.2, and which may be of use for various groups G to show that $C^*_\lambda(G)$ is simple. (We note that in [3, 17] other new tests are given of different natures, which do not appear to directly target the issue we are discussing here.)

In a fashion similar to Condition 4.2, our new Condition 4.5 gives the simplicity of $C^*_\lambda(G)$ for a group G if for all finite subsets H of $G\backslash\{e\}$, we can carry out a certain construction (creating large free subgroups in a certain way from H). The new condition is as follows.

Condition 4.5 *For all finite non-empty subsets $H \subset G$, $e \notin H$ and for all positive integers n there are r, $s \in G$ and a set $\{c_1, c_2, \ldots, c_n\} \subset C_G(sr)$ such that the set $\{(sgr)^{c_k} : k = 1, \ldots, n\}$ is free for all $g \in H$.*

Of course by taking r and s above to be the identity, we see immediately that anytime a group satisfies Condition 4.2, it also satisfies Condition 4.5.

Unfortunately, Condition 4.5 is still not weak enough to show that the algebra $C^*_\lambda(T)$ is simple. And, as in the previous case, this is still a consequence of the fact that R. Thompson's group F admits no free subgroups, although we have to work harder to break Condition 4.5. In particular, we observe that there is no particular obstruction stopping a discrete group $G \leq Homeo_+(\mathbb{S}^1)$, where G admits subgroups H with H admitting a global fixed point on the circle and also free, non-abelian subgroups, from satisfying Condition 4.5.

Indeed, such groups are rampant in $Homeo_+(\mathbb{S}^1)$, and it is quite plausible that some such groups will satisfy Condition 4.5, while also failing Condition 4.2.

We also mention the new paper [1], written after the first draft of this chapter, which shows that if F is amenable, then the C^* algebra of T cannot be simple, as T admits F as a *normalish* subgroup, so that in this case the amenable radical of T is not trivial (following a method of [3].)

4.1.2 On Amenability

There has also been a long-standing interest in the question of the amenability of Richard Thompson's group F, introduced in Thompson's notes of 1965 (see the survey [7] for a general background on the three Thompson groups $F < T < V$), with many failed attempts to prove either the amenability or non-amenability of F. The groups $F < T < V$ arise in many areas of mathematics for reasons which are not entirely understood. One plausible explanation is that they express in some fundamental way connections through Category Theory with associativity and versions of commutativity (see [4, 10, 8] for some discussion of these connections), which of course are fundamental aspects of any theory involving products. Regardless of the cause, it is still the case that these groups arise naturally in many areas of mathematics including dynamics, logic, topology, and more obviously geometric group theory. One fetching example of such an appearance is in, e.g., the relationship between the group F and the theory of associahedra, and in particular, the theory related to the the proof of the Four Color Theorem [5]. In any case, it is well known now that the Richard Thompson groups are fundamental. One of the initial motivations for the work in this chapter was the inspiration given by the work of Haagerup and Olesen in [12].

Uffe Haagerup and Kristian Knudsen Olesen in [12] (see also [20, Section 4.2]) show that the simplicity of $C^*_\lambda(T)$ implies the non-amenability of F via a construction given below (see Proposition 4.19 below). Observe that it is well known that T is a group with unique trace. Thus, our initial motivation in this chapter was to investigate the ideal structure of $C^*_\lambda(T)$. And indeed, we find we can give a partial converse to the Haagerup–Olesen result, which is of interest, although it is not hard to show.

Haagerup and Olesen's idea showing that the simplicity of the algebra $C^*_\lambda(T)$ implies the non-amenability of F runs as follows. Consider T "acting" on the interval $[0, 1]$. Assume that the stabiliser of 0, which is the standard copy of the Thompson group F in T, is amenable. Since the action of T on $\mathbb{Z}[\frac{1}{2}]$ is transitive, we have that the representation induced by this action, $\pi : T \to B(l_2(\mathbb{Z}[\frac{1}{2}]))$, is weakly contained in the left regular representation. From this one sees that there is a unique $*$-homomorphism from $C^*_\lambda(T)$ into the C^*-algebra generated by $\pi(T)$.

Consider now a finite subset H of T so that $H = H_1 \sqcup H_2$ and with

$$\sum_{g\in H_1} \pi(g) - \sum_{h\in H_2} \pi(h) = 0.$$

The simplicity of $C^*_\lambda(T)$ now implies that the ideal generated by $\sum_{g\in H_1} \lambda(g) - \sum_{h\in H_2} \lambda(g)$ is proper. However, this is not possible since π is non-trivial, so F must be non-amenable.

As a converse of the above we have the following. Thompson's group

F is non-amenable if and only if there exists a finite set H which can be decomposed as a disjoint union of sets H_1 and H_2 with $\sum_{g\in H_1}\pi(g) = \sum_{h\in H_2}\pi(h)$ and such that the closed ideal generated by $\sum_{g\in H_1}\lambda(g) - \sum_{h\in H_2}\lambda(h)$ coincides with $C^*_\lambda(T)$.

4.1.3 On Free Subgroups

In the discussions in the last two subsections, we see that finding free subgroups is an important part of satisfying either of the core conditions detecting simplicity, and so we find tools to do just that. The results we mention here are somewhat specialized, but they may be of interest to researchers working to find free subgroups in groups of homeomorphisms of the circle in the first case, and in the second case, to researchers working to find particular free subgroups in R. Thompson's group T.

Our first proposition (along these lines) detects a free basis in a group G of homeomorphisms of the circle, where G is generated by elements with rational rotation number. The main value of the proposition is that it finds the free basis using essentially the element given, if they satisfy the correct initial conditions.

Proposition 4.6 *Suppose a and b are orientation preserving homeomorphisms of the circle S^1 with rational rotation numbers $Rot(a) = p/q$ and $Rot(b) = r/s$ in lowest non-negative terms, where*

1 b is not torsion, and
2 if $x \in Fix(b^s)$ and $j \in \mathbb{Z}$ with $a^j \neq 1_T$ are given, then we have $a^j x \notin Fix(b^s)$.

In these circumstances, there is a positive integer k so that a and b^k are a free basis for the group $\langle a, b^k\rangle$.

The next lemma shows that one can use a set of elements with a common point of support to make large rank free groups R. Thompson's group T in such a way as to satisfy our Condition 4.5.

Lemma 4.7 *Let H be a finite set of non-trivial elements in T so that there is some point $p \in \cap_{h\in H} Supp(h)$. Then, for any positive integer n there is an element $g \in T$ and $\{c_1, c_2, \ldots, c_n\}$ so that $c_i \in C_T(g)$ for all i, and so that for all $h \in H$ we have that the set*

$$G_h := \{(gh)^{c_i} \mid i \in \{1, 2, \ldots, n\}\}$$

is a free basis for a free group of rank n.

And finally, we show that if we have any finite set H of non-trivial elements in T, then we can very nearly satisfy our Condition 4.5; we need only allow passing to a fixed power p of the elements $(hg)^{c_i}$, for $p = |H|$, to find a free basis of rank n. Note again that this cannot happen for the original Kesten Condition 4.2.

Lemma 4.8 *Let H be a finite set of non-trivial elements in T with cardinality p. Then there is an element $g \in T$ such that for any positive integer n there are elements $\{c_1, c_2, \ldots, c_n\}$ so that $c_i \in C_T(g)$ for all i, and so that for all $h \in H$ the set*

$$G_h := \{((hg)^{c_i})^p \mid i \in \{1, 2, \ldots, n\}\}$$

freely generates a free group of rank n.

4.1.4 A Description of What Comes Next

In Section 4.3 we provide a short discussion of the historical Ping-Pong Lemma and we prove a version of the Ping-Pong Lemma (Proposition 4.6) useful for detecting free subgroups when considering a group generated by two homeomorphisms of S^1 with rational rotation numbers. Proposition 4.6 is an essential ingredient in the proof of Lemma 4.7.

In Section 4.4, we give proofs that we can carry out the construction of Condition 4.5 for many cases of finite $H \subset T \setminus \{e\}$, including the cases we need to prove Lemma 4.7. We also describe some cases of H where we cannot carry out the construction of Condition 4.5, but where related constructions do produce large free subgroups.

In Section 4.5 we present our proof of the second result, modulo our Lemma 4.7.

In our final Section 4.6 we state some remaining questions which we find interesting.

4.2 Powers' Test

We now prove Lemma 4.3. Recall the statement.

Lemma 4.3 (restated) *Let X be a set, and G the group of bijections from X to X. Suppose h_1, $h_2 \in G \setminus \{1\}$ so that $Supp(h_1) \cap Supp(h_2) = \emptyset$. If c_1, $c_2 \in G$ so that $Supp(h_1^{c_1}) \cup Supp(h_1^{c_2}) = X$ then*

$$Supp(h_2^{c_1}) \cap Supp(h_2^{c_2}) = \emptyset.$$

Proof Suppose

$$X = Supp(h_1^{c_1}) \cup Supp(h_1^{c_2})(= c_1 \cdot Supp(h_1) \cup c_2 \cdot Supp(h_1)).$$

If there is an $x \in X$ so that $x \in Supp(h_2^{c_1}) \cap Supp(h_2^{c_2})$, then $x = c_1 \cdot y$ and $x = c_2 \cdot z$, where y and z are in $Fix(h_1)$. In particular, $x \in c_1 \cdot Fix(h_1) \cap c_2 \cdot Fix(h_1)$. This implies that $c_1 \cdot Supp(h_1) \cup c_2 \cdot Supp(h_1) \neq X$. □

As mentioned before, Lemma 4.3 immediately implies that we cannot use Condition 4.2 when approaching the question of the simplicity of the algebra $C^*_\lambda(T)$.

Corollary 4.9 *Suppose that $H \subset T$ admits elements h_1 and h_2 so that $Supp(h_1) \cap Supp(h_2) = \emptyset$. Then for $n \geq 2$ there is no set of elements $\{c_1, c_2 \ldots, c_n\}$ so that $\langle h^{c_1}, h^{c_2}, \ldots, h^{c_n} \rangle$ is a free group on n generators for all $h \in H$.*

Proof Suppose $H := \{h_1, h_2, \ldots, h_k\}$ is a finite set with cardinality at least two, and h_1 and h_2 are in H so that $Supp(h_1) \cap Supp(h_2) = \emptyset$. Further suppose that $n \geq 2$ is fixed and $c_1, c_2, \ldots, c_n$ are chosen so that for all $h \in H$, we have that $\langle h^{c_1}, h^{c_2}, \ldots, h^{c_n} \rangle$ is free on n generators. As proven in Brin and Squier's paper [6], the group of piecewise linear homeomorphisms of the unit interval has no non-abelian free subgroups, so we see immediately that $Supp(h_1^{c_1}) \cup Supp(h_1^{c_2}) = S^1$. Now by Lemma 4.3 we know that $Supp(h_2^{c_1}) \cap Supp(h_2^{c_2}) = \emptyset$. Therefore $\langle h_2^{c_1}, h_2^{c_2} \rangle \cong \mathbb{Z} \times \mathbb{Z}$. □

We now carry out some work in order to offer an apparently weaker version of Condition 4.2 which will be used throughout the remainder of this article. In particular, we need the supporting theorem below.

Theorem 4.10 *Let $H \subset G$ be a finite set and assume there is an element $w \in H$ such that for all positive integers n there is a set $\{c_1, c_2, \ldots, c_n\} \subset G$ and $r, s \in G$ such that $c_i \in C_G(swr)$ for all i and*

$$\lim_{n \to \infty} \frac{1}{n} ||\Sigma_{i=1}^n \lambda(c_i sgr {c_i}^{-1})|| = 0, \text{ for all } g \in H \backslash \{w\},$$

then for all coefficients β_g indexed by H with $\beta_w \neq 0$, the ideal generated by $\sum_{g \in H} \beta_g \lambda(g)$ is equal to $C^*_\lambda(G)$.

Proof Let I be an ideal in $C^*_\lambda(G)$ generated by $b := \sum_{g \in H} \beta_g \lambda(g)$. Assume that I is proper. The closure of I is proper, thus we can assume

I is closed. Note that $\Sigma_{i=1}^{n}\lambda(c_i s)b\lambda(rc_i^{-1}) \in I$. Since $c_i \in C_G(swr)$ we have

$$\begin{aligned}\|\lambda(swr)-&\frac{1}{\beta_w n}\Sigma_{i=1}^{n}\lambda(c_i s)b\lambda(rc_i^{-1})\|\\ =&\frac{1}{\beta_w n}\|\Sigma_{g\in H\setminus\{w\}}\Sigma_{i=1}^{n}\beta_g\lambda(c_i sgrc_i^{-1})\|\\ \le&\frac{1}{\beta_w}\Sigma_{g\in H\setminus\{w\}}|\beta_g|\frac{1}{n}\|\Sigma_{i=1}^{n}\lambda(c_i sgrc_i^{-1})\|\\ =&\max\{|\beta_g|/\beta_w : g\in H\}\cdot\max\left\{\frac{1}{n}\|\Sigma_{i=1}^{n}\lambda(c_i sgrc_i^{-1})\| : g\in H\right\}.\end{aligned}$$

By our assumptions, the last quantity can be arbitrarily small for large n. Thus there is an element in I which is at distance less than 1 from a unitary operator, which implies that it is invertible and $I = C^*_\lambda(G)$. □

Applying the theorem above to the set $H \cup \{e\}$ shows that the following condition implies simplicity of $C^*_\lambda(G)$.

Condition 4.5 (restated) *For all finite non-empty subsets $H \subset G$, $e \notin H$ and for all positive integers n there are r, $s \in G$ and a set $\{c_1, c_2, \ldots, c_n\} \subset C_G(sr)$ such that the set $\{c_k(sgr)c_k^{-1} : k = 1, \ldots, n\}$ is free for all $g \in H$.*

Condition 4.2 implies Condition 4.5. However, Condition 4.5 is still inadequate for showing that $C^*_\lambda(T)$ is simple.

Lemma 4.11 *There are elements g_1, $g_2 \in T\setminus\{e\}$ with the property that, for any r, $s \in T$, there are no elements c_1, c_2, c_3, and $c_4 \in C_T(sr)$ with both subgroups*

$$G_1 = \langle (sg_1r)^{c_1}, (sg_1r)^{c_2}, (sg_1r)^{c_3}, (sg_1r)^{c_4}\rangle, \text{ and}$$

$$G_2 = \langle (sg_2r)^{c_1}, (sg_2r)^{c_2}, (sg_2r)^{c_3}, (sg_2r)^{c_4}\rangle$$

free on four generators.

Proof Let g_1, $g_2 \in T$ so that $Supp(g_1) = (0, 1/2)$ and $Supp(g_2) = (1/2, 1)$. Let r and $s \in T$ and suppose c_1, c_2, c_3 and $c_4 \in C_T(sr)$. Set $k_{ij} = (sg_ir)^{c_j}$, for i, $j \in \{1, 2, 3, 4\}$, and suppose that c_1, c_2, c_3 and c_4 were chosen so that both groups $G_i = \langle k_{i1}, k_{i2}, k_{i3}, k_{i4}\rangle$ are free on four generators for $1 \le i \le 2$.

Consider the intervals $X_{i1} = (c_1r^{-1})\cdot Fix(g_i)$, $X_{i2} = (c_2r^{-1})\cdot Fix(g_i)$, $X_{i3} = (c_3r^{-1})\cdot Fix(g_i)$, and $X_{i4} = (c_4r^{-1})\cdot Fix(g_i)$. If $x_{ij} \in X_{ij}$, then $k_{ij}\cdot x_{ij} = c_jsg_irc_j^{-1}\cdot x_{ij} = (c_jsg_i)\cdot y_{ij} = (c_js)\cdot y_{ij} = (c_jsrc_j^{-1})\cdot x_{ij} =$

$(sr) \cdot x_{ij}$, as for all i, j we have $y_{ij} \in Fix(g_i)$. That is, k_{ij} acts as sr over X_{ij}.

Further, consider the elements $f_{i,ab} = k_{ia}^{-1} k_{ib}$, where $i \in \{1,2\}$ and $a \neq b \in \{1,2,3,4\}$. It is immediate that $\langle f_{i,ab}, f_{i,cd} \rangle$ is free on two generators if either $b \neq c$ or $d \neq a$. Therefore, by Brin and Squier's result (from [6]) that $PL_o(I)$ has no non-abelian free subgroups, we know that $Fix(f_{i,ab}) \cap Fix(f_{i,cd}) = \emptyset$ for $i \in \{1,2\}$ and either $b \neq c$ or $d \neq a$. Now, for instance, if there is an index i and some point $p \in X_{i1} \cap X_{i2} \cap X_{i3}$, then both $f_{i,12}$ and $f_{i,13}$ must fix p, which is a contradiction. Therefore we see that X_{i1}, X_{i2}, and X_{i3} cannot share a common point for any index i. By the same argument, for any valid indices i, a, b, and c (where $i \in \{1,2\}$ and $a \neq b \neq c \neq a$) we see that $X_{ia} \cap X_{ib} \cap X_{ic} = \emptyset$.

One now sees immediately that for any valid indices i, a, b, and c (where $i \in \{1,2\}$ and $a \neq b \neq c \neq a$) we must also have that $X_{ia} \cup X_{ib} \cup X_{ic} = S^1$. This follows as otherwise there is some point p in the intersection $X_{ja} \cap X_{jb} \cap X_{jc}$ for the index $j \neq i$ (since $X_{1*} = \overline{S^1 \backslash X_{2*}}$ for any index $*$).

Suppose that for some indices i, $a \neq b$ we have that $X_{ia} \subset X_{ib}$, and let c and d be the two remaining distinct indices of $\{1,2,3,4\} \backslash \{a,b\}$. Let p be an endpoint of X_{ib}. We have that p must be in both X_{ic} and X_{id}, otherwise there will be some point $q \in S^1 \backslash X_{ib}$ which is near to p so that q is not in either of $X_{ia} \cup X_{ib} \cup X_{ic} = X_{ib} \cup X_{ic}$ or $X_{ia} \cup X_{ib} \cup X_{id} = X_{ib} \cup X_{id}$. But this contradicts the fact that $X_{ib} \cap X_{ic} \cap X_{id} = \emptyset$.

It now immediately follows that for any index i and two distinct indices a and b, we have that $X_{ia} \cap X_{ib}$ is a non-empty closed interval (possibly a single point) while $X_{ia} \cup X_{ib}$ is also a closed interval which misses some points in S^1.

But now we are done, as follows. For any index i the intervals X_{i1}, X_{i2}, and X_{i3} cover the circle, and have the properties that each pair of sets intersects in an interval, and no pair covers the whole circle. Now consider X_{i4}. It must likewise intersect both X_{i1} and X_{i2} non-trivially, and the union of X_{i1}, X_{i2}, and X_{i4} also covers the whole circle. Therefore the end of X_{i1} which is not in X_{i2}, is in both X_{i3} and X_{i4}. Hence $X_{i1} \cap X_{i3} \cap X_{i4} \neq \emptyset$, which implies that the group G_i cannot be free on four generators, as $f_{i,13}$ and $f_{i,14}$ share a common fixed point and will not generate a free subgroup of G_i. □

Remark 4.12 We observe that it is still plausible that even with g_1 and g_2 as in the proof above (supports over $(0, 1/2)$ and $(1/2, 1)$, respectively), one could potentially find r, s, and c_1, c_2, and $c_3 \in C_T(sr)$ so that setting $k_{ij} = c_j s g_i r c_j^{-1}$ as above we would have $H_r = \langle k_{r1}, k_{r2}, k_{r3} \rangle$ free on three generators for both $r = 1$ and $r = 2$. The corresponding claim for even two generator free groups could not be conceived of under Condition 4.2.

4.3 A Ping-Pong Lemma for Orientation Preserving Homeomorphisms of S^1

In this section, we prove a version of the Ping-Pong Lemma which we are using in our main argument. In the notations below we write all actions as left actions, in keeping with the tradition in the C^* literature, although much of the Thompson groups literature uses right action. In particular, if $x \in S^1$ and $s, t \in T$, we write tx for the image of x under t, and the conjugation $s^t := sts^{-1}$, which means: apply s^{-1} first, then t, and then s. We consider finite sets with repetitions.

In support of that lemma we ask the reader to recall an ordinary statement of Fricke and Klein's Ping-Pong Lemma (first proven in [9], but we give a different statement), and two further facts, one quite classical.

Lemma 4.13 *(Ping-Pong Lemma) Let G be a group of permutations on a set X, and let $a, b \in G$, where $b^2 \neq 1$. If X_a and X_b are two subsets of X so that neither is contained in the other, and for all integers n we have $b^n \cdot X_a \subset X_b$ whenever $b^n \neq 1$, and $a^n \cdot X_b \subset X_a$ whenever $a^n \neq 1$, then $\langle a, b \rangle$ factors naturally as the free product of $\langle a \rangle$ and $\langle b \rangle$. In particular, $\langle a, b \rangle \cong \langle a \rangle * \langle b \rangle$.*

Suppose that $f : S^1 \to S^1$ is an orientation preserving homeomorphism of the circle $S^1 = \mathbb{R}/\mathbb{Z}$, then f may be lifted to a homeomorphism of $\mathbb{R}$ by $F(x + m) = F(x) + m$ for every x and m. The rotation number of f is defined to be $Rot(f) = \lim_{n \to \infty}(F^n(x) - x)/n$. The following theorem is generally relevant to the arguments in the final section of this chapter, and appears first in [11], although there now exist many different proofs, the shortest of which appears to be in [2].

Theorem 4.14 *Every element of Thompson's group T has rational rotation number.*

The last tool we need in order to establish our own version of the Ping-Pong Lemma is the following classical result of Poincaré.

Lemma 4.15 *(Poincaré's Lemma, circa 1905) If f is an orientation* *preserving homeomorphism of S^1 and f has rotation number p/q in lowest terms, then there is an orbit in S^1 of size exactly q under the action of $\langle f \rangle$.*

We are now in a good position to quote and prove our main technical tool.

Proposition 4.16 *Suppose a and b are orientation preserving homeomorphisms of the circle S^1 with rational rotation numbers $Rot(a) = p/q$ and $Rot(b) = r/s$ in lowest non-negative terms, where*

1 b is not torsion, and
2 if $x \in Fix(b^s)$ and $j \in \mathbb{Z}$ with $a^j \neq 1_T$ are given, then we have $a^j x \notin Fix(b^s)$.

In these circumstances, there is a positive integer k so that a and b^k are a free basis for the group $\langle a, b^k \rangle$.

Proof In the proof below, let us take a, $b \in T$ and p, q, r, $s \in \mathbb{N}$ as in the statement of the lemma. Set $b_0 := b$. We will occasionally update to a new version of b, which will be given by a new index. The new b will always be an integral power of the previous indexed b.

Set $b_1 := b_0^s$. The element b_1 will have rotation number $0/1$ in lowest non-negative terms. For b_1, we have that $Fix(b_1)$ is not empty, and also not the whole circle (otherwise b was originally a torsion element in T).

Let $\mathscr{I} \subset S^1$ be such that for each component C of $Supp(b_1)$, we have $|C \cap \mathscr{I}| = 1$, and associate each such C with its unique point in $\mathscr{I}$, so that $\mathscr{I}$ becomes an index set for the components of $Supp(b_1)$. We observe that $\mathscr{I}$ comes with an inherent circular order as a subset of S^1. Let L_b represent the set of limit points of $\mathscr{I}$ which are not in $\mathscr{I}$, and observe that $L_b \subset Fix(b_1)$.

For each positive integer d, set $\Delta_d := [-d, d] \cap (\mathbb{Z}\backslash\{0\})$, the set of non-zero integers a distance d or less from zero. Now for all positive integers d we can set ϵ_d to be one half of the distance from $Fix(b_1)$ to the set $\cup_{i \in \Delta_d} a^i \cdot Fix(b_1)$. Noting that these ϵ_d are all well defined and non-zero (unless a is torsion) as the sets involved are compact and as $a^m \cdot Fix(b_1) \cap a^n \cdot Fix(b_1) \neq \emptyset$ implies that either $m = n$, or a is torsion and $n - m$ is divisible by the order of a.

Our analysis now splits, depending on whether or not a is torsion. In the case that a is torsion, our proof is somewhat easier, so we will execute that proof immediately.

Case: a is torsion with order q

In this case, the value ϵ_{q-1} explicitly measures one half of the distance between $Fix(b_1)$ and the union of the images of $Fix(b_1)$ under the action of non-trivial powers of a. Set $\mathcal{U}$ to be the open ϵ_{q-1} neighbourhood of $Fix(b_1)$, and observe that for each integer $i \in \{1, 2, \ldots, q-1\}$ we have $a^i \cdot Fix(b_1) \cap \mathcal{U} = \emptyset$. For each non-zero $i \in \{1, 2, \ldots q-1\}$ set $\mathcal{U}_i$ to be the ϵ_{q-1} neighbourhood of $a^i \cdot Fix(b)$. Again, for all such indices i, $\mathcal{U} \cap \mathcal{U}_i = \emptyset$. Set

$$X_b := \mathcal{U} \bigcap_{1 \leq i < q} (a^{q-i} \cdot \mathcal{U}_i).$$

Now by construction we have that the image set $a^i \cdot X_b \subset \mathcal{U}_i$, but $\mathcal{U}_i \cap X_b = \emptyset$ since $\mathcal{U}_i \cap \mathcal{U} = \emptyset$ and $X_b \subset \mathcal{U}$. Thus, X_b is an open set containing $Fix(b_1)$ that is disjoint from its image under the action of any non-trivial power of a.

As X_b and the components of support of b_1 altogether cover the circle, there is a finite set of open interval components of $Supp(b_1)$ which together with X_b covers the circle. In turn, this implies there is a minimal positive integer v so that for all $x \in S^1 \backslash X_b$, we have $b_1^v x \in X_b$ and $b_1^{-v} x \in X_b$.

Now we can set

$$X_a := \cup_{1 \leq i < q} (a^i \cdot X_b).$$

With this choice of X_a and X_b we have arranged that we satisfy the hypotheses of Lemma 4.13 for the elements a and a^k where $k = v \cdot s$.

Case: a is not torsion

Throughout this case, given a set $X \subset S^1$, and $\epsilon > 0$, we shall use the notation $N_\epsilon(X)$ to denote the open ϵ-neighbourhood of X, that is, all points in S^1 a distance less than ϵ from some point in X.

In this case with a not torsion, we must specify the set $F_a := Fix(a^q)$, which is a closed non-empty subset of the circle which is disjoint from $Fix(b_1)$. Choose a specific $\epsilon > 0$ so that $N_\epsilon(Fix(b_1)) \cap N_\epsilon(F_a) = \emptyset$, noting that such an epsilon value exists as $Fix(b_1)$ and F_a are disjoint compact subsets of S^1.

Let m be a positive integer so that both $a^{mq} \cdot Fix(b_1) \subset N_\epsilon(F_a)$ and $a^{-mq} \cdot Fix(b_1) \subset N_\epsilon(F_a)$. This m exists as a^q acts as a monotone strictly increasing, or as a monotone strictly decreasing function over each component of its support, and as the limit point of any point in a component of support of a^q under increasing powers of a^q must be a fixed point of a^q (and similarly under negative powers of a^q), and as $Fix(b_1)$ is a compact set and hence is contained in a union of finitely many components of support of a^q.

We now observe that for n an integer with $|n| > m$, we have that $a^{nq} \cdot Fix(b_1) \subset F_a$ as well. We would like to argue a stronger result now that there is a positive constant N so that for all $j > N$ we have $a^j \cdot Fix(b_1) \subset N_\epsilon(F_a)$ and $a^{-j} \cdot Fix(b_1) \subset N_\epsilon(F_a)$.

To make this argument, the main point to observe is that there is an induced action of $\langle a \rangle$ on the set of components of support of a^q which partitions these components into (possibly infinitely many) orbits of size q. Further, as a commutes with a^q and a is orientation preserving, it is easy to see that each such orbit consists of components of support where the action of a^q is increasing on all components of the orbit, or decreasing on all components of the orbit.

It is also the case that there are only finitely many components of support of a^q which are not already wholly contained in $N_\epsilon(F_a)$. Let $C_1, C_2, \ldots, C_w$ represent these components, and observe that $Fix(b_1)$ is contained in the union K of these compact intervals. For each component C_j, let I_j be the closed interval $C_j \backslash N_\epsilon(F_a)$. Now each of these components C_j are in an orbit of length q amongst the components of support of a^q, and in each such orbit the action of a^q on each component is in the same direction. Hence there is a finite number N so that for all $j > N$ and intervals I_m, we have that $a^j \cdot I_m \subset N_\epsilon(F_a)$, and also $a^{-j} \cdot I_m \subset N_\epsilon(F_a)$.

Now define J to be

$$J := \cup_{i \in \Delta_N}((a^i \cdot Fix(b_1)) \cap K),$$

where we recall that $\Delta_n = [-n, n] \cap (\mathbb{Z} \backslash \{0\})$ for any particular $n \in \mathbb{N}$.

It is immediate that J is a compact set which is disjoint from $Fix(b_1)$. As such, there is a $\delta > 0$ so that $\delta < \epsilon$ and the δ-neighbourhood $N_\delta(Fix(b_1))$ of $Fix(b_1)$ is disjoint from the set V_δ defined as

$$V_\delta := \cup_{i \in \Delta_N}(a^i \cdot N_\delta(Fix(b_1)))$$

and noting that as $\delta < \epsilon$ we also have that $N_\delta(Fix(b_1))$ is disjoint from $N_\epsilon(F_a)$.

Now set $X_b := N_\delta(Fix(b_1))$ and $X_a := N_\epsilon(F_a) \cup V_\delta$.

By construction, there is an integer $z > 0$ so that b_1^z takes the complement of X_b (and so, X_a) into X_b, while all non-trivial powers of a take X_b into X_a. Hence the integer $k = s \cdot z$ has the property that a and b^k freely generate a free group of rank 2. □

4.4 Applying Condition 4.5, and Variants, in T

Here we list lemmas where Condition 4.5 can be used.

Lemma 4.17 *Let H be a finite set of non-trivial elements in T so that there is some point $p \in \cap_{h \in H} Supp(h)$. Then, for any positive integer n there is an element $g \in T$ and $\{c_1, c_2, \ldots, c_n\}$ so that $c_i \in C_T(g)$ for all i, and so that for all $h \in H$ we have that the set*

$$G_h := \{(gh)^{c_i} \mid i \in \{1, 2, \ldots, n\}\}$$

is a free basis for a free group of rank n.

Proof Let H and p be as in the statement of the lemma, and let $n \in \mathbb{N}$ be given. For each $h \in H$, let $Rot(h) := r_h/s_h$ written in lowest terms (N.B.: any finite periodic orbit under the action of $\langle h \rangle$ is of length s_h). By the definition of p, we see there is a non-empty interval (a, b) with $p \in (a, b)$ so that for all $h \in H$ we have $(a, b) \cdot h^j \cap (a, b) = \emptyset$ for all $1 \leq j < s_h$.

Now let $g \in T$ be an element with rotation number $Rot(g) = 0$ which fixes exactly the point p. We can choose g so that for all $h \in H$ the product gh has precisely two fixed points which are generated from the contact of the graph of g with the graph of the element h^{-1}, by choosing the graph of g to be near to a step function (the nearly vertical component of the graph of g should be steeper than any slope of any element of H, and the nearly horizontal component of the graph of g should have slope closer to zero than any slope of any element of H, and, in a small epsilon box around (p, p) which misses the graphs of the elements of H, the graph of g is unrestrained). As g fixes the point p, it is the case that for all elements $h \in H$, we have $k_h \cdot p = gh \cdot p = g \cdot (h \cdot p) \neq p$ and further that under the action of $\langle k_h \rangle$, p is in an infinite orbit which limits to the two ends of the component of support of k_h which contains p. In these conditions we can immediately apply Proposition 4.6 to claim that for each $h \in H$, there is a power g^{ρ_h} of g so that k_h and the element g^{ρ_h} freely generate a free group of rank 2. Now take $\theta := LCM\{\rho_h \mid h \in H\}$, so that g^θ is free for k_h for each h.

Now, it is immediate that setting $c_i := g^{i \cdot \theta}$ for indices i in the range $1 \leq i \leq n$, we can create sets $X_h := \{(gh)^{c_i}\}_{i=1}^{n}$ so that $\langle X_h \rangle$ is free on n-generators for all $h \in H$.

□

The following is some weakening of Condition 4.5 which we can achieve in Thompson's group T. Unfortunately, by raising to the power p, we find a free subgroup of $K := \{((hg)^{c_i}) \mid i \in \{1,2,\ldots,n\}\}$ which is generally not finite index.

Lemma 4.18 *Let H be a finite set of non-trivial elements in T with cardinality p. Then there is an element $g \in T$ such that for any positive integer n there are elements $\{c_1, c_2, \ldots, c_n\}$ so that $c_i \in C_T(g)$ for all i, and so that for all $h \in H$ the set*

$$G_h := \{((hg)^{c_i})^p \mid i \in \{1,2,\ldots,n\}\}$$

freely generates a free group of rank n.

Proof Index the $h \in H$ so that $H = \{h_1, h_2, \ldots, h_p\}$ for some minimal positive integer p. As each element $t \in T$ has a rational rotation number $Rot(t) \in \mathbb{Q}/\mathbb{Z}$, let $Rot(h_i) = r_i/s_i$ expressed in lowest positive terms. For each index i, inductively choose a point $x_i \in Supp(h_i)$ so that the set

$$X_i := \left\{h_j^{-1} \cdot x_i, x_i, h_j \cdot x_i \mid 1 \leq j \leq i\right\}$$

is disjoint from the union $\cup_{j<i} X_j$, and where each x_i is chosen as a dyadic rational so that no point in X_i is an end of a component of support of $h_j^{s_j}$ for any index j. Note that in all cases we are choosing an x_i so that a finite set of images miss a finite set of points in the circle, and this is easy to do. Set $P := \{x_i \mid 1 \leq i \leq p\}$. Re-index P so that $0 \leq x_1 < x_2 < \cdots < x_p < 1$, and re-index the sets X_i correspondingly.

Now choose, for each i, an interval (a_i, b_i) centred around x_i, and set

$$I_i := \left\{h_j^{-1} \cdot (a_i, b_i), (a_i, b_i), h_j \cdot (a_i, b_i) \mid 1 \leq j \leq i\right\},$$

where each (a_i, b_i) is chosen small enough so that each element of I_i is disjoint from the union $\cup_{j<i, B \in I_j} B$, and where (a_i, b_i) is disjoint from $h_j^{-1} \cdot (a_i, b_i)$ and $h_j \cdot (a_i, b_i)$ whenever $x_i \in Supp(h_j)$, and where $h_j^{-1} \cdot (a_i, b_i)$ and $h_j \cdot (a_i, b_i)$ intersect each other non-trivially only if h_j is torsion of order two. We can insist these intervals are possibly smaller still, so that the complement of the union $\tilde{U} := \cup_{1 \leq j \leq p, B \in I_j} B$ is a union of closed intervals each of which has non-empty interior. We observe that by our choices of the sets X_i, we can so choose our intervals (a_i, b_i).

Now let $\tilde{g}$ be a non-torsion element with rotation number $1/p$ which admits exactly one orbit of size p under the action of $\langle \tilde{g} \rangle$, where the orbit is the set P, and where $\tilde{g}$ cyclically permutes the x_i in order, taking x_1 to x_2, x_2 to x_3, etc.

The element $\gamma := \tilde{g}^p$ has p components of support $S_1 := (x_1, x_2)$, $S_2 := (x_2, x_3)$, $\ldots$, $S_p := (x_p, x_1)$, and γ is either increasing on each component of support, or decreasing on each component of support. We re-choose $\tilde{g}$ if necessary so that it satisfies all previous conditions and so that γ is increasing on each component of support, and note that such elements $\tilde{g}$ exist.

Now set $U := \cup_{1\leq j\leq p}(a_i, b_i)$, which union also has the property that the complement $C := S^1\backslash U$ is a union of p disjoint closed intervals $D_i := [b_i, a_{i+1}]$ (where we set $a_{p+1} := a_1$) each of which is not a point. As such is the case, we can find a positive integer d so that $\gamma^d \cdot C \subset U$ as after d iterations of γ, the set D_i will be moved so as to have image just to the left of x_{i+1} (where again, we set $x_{p+1} := x_1$).

Now set $g := \gamma^d\tilde{g}$. The element g so constructed has rotation number $1/p$ as before, and acts on the set P as $\tilde{g}$ does, but it has the further property that for all integers $s \neq 0$, $g^s \cdot C \subset U$.

Now set, for each index i, $k_i := gh_i$. We will now show that for all k_i, the point x_j is not in any finite periodic orbit of k_i, for any index j.

Let i and j be indices so that k_i and x_j are an element and point as defined above, respectively. Let q be an index and consider the interval $(a_q, x_q]$ under the action of g, and then consider the interval $(a_j, x_j]$ under the action of k_i.

Under the action of g, we know already that $x_q \mapsto x_{q+1}$, and that the component $[b_{q-1}, a_q]$ of C maps into the interval (a_{q+1}, x_{q+1}). In particular, the whole interval $[b_{q-1}, x_q]$ is mapped into $(a_{q+1}, x_{q+1}]$ by g and so the interval $[a_q, x_q]$ is mapped into $(a_{q+1}, x_{q+1}]$ as well, and precisely the point x_q will map to x_{q+1}.

Now consider the action of k_i on $(a_j, x_j]$. If x_j is in the support of h_i, then $h_i \cdot (a_j, x_j] \subset C$, whereupon that image is moved into some interval (a_*, x_*) by the action of g.

In particular, $k_i^z \cdot x_j = x_{j+z}$ (in cyclic order) as long as for all $0 \leq y \leq z$ we have $x_{j+y} \notin Supp(h_i)$. If instead for some minimal integer $s \in \{0, 1, \ldots, p-1\}$ we have $x_{j+s} \in Supp(h_i)$ then k_i^{s+1} will move x_j out of P as we will have $k_i^{s+1} \cdot x_j \in (a_*, x_*)$ for some index $*$. Furthermore, if this happens, all further iterates of x_j will fail to re-enter P.

However, x_i is itself in the support of h_i, so this eventuality happens (in p steps or fewer), and so no x_j is on a finite periodic orbit under the action of $\langle k_i\rangle$, for any indices j and i.

In particular, for all indices j, we see that $(k_j^{tp} \cdot P) \cap P = \emptyset$ for all $t \in \mathbb{Z}\backslash \{0\}$.

We can now quote Proposition 4.6, using $a := k_j^p$ for each index j and $b = g$ to claim that for each index j there is an integer z_j so that the group $\langle k_i^p, g^{z_j}\rangle$ is free on two generators (in this setup, $P = Fix(b^s) =$

$Fix(g^p)$). Now set $z := LCM(\{z_1, z_2, \ldots, z_p\})$, the least common multiple of the values z_j, and then we have that for all indices j, the elements k_j^p and g^z generate a free group. In particular, setting $c_i := g^{iz}$ for $i \in \{1, 2, \ldots, n\}$ we see that $\Gamma_j := \{(k_j^p)^{c_i} = (k_j^{c_i})^p \mid 1 \leq i \leq n, i \in \mathbb{Z}\}$ freely generates a free group of rank n.

But now, recalling that $k_i = gh_i$, and that therefore $[g, c_i] = 1$, we have proved our claim. □

4.5 Non-amenability of F and a Condition on Ideals of its C^*-algebra.

In this section we will show the following, which is the converse of the Haagerup-Olesen result. The proof can be done in several ways. One is to use Lemma 4.7 and follow Powers' proof of simplicity, another is to show it directly using the regular Ping-Pong Lemma. This happens because the generators can be chosen to act in the same way on some subinterval of $[0, 1]$.

Proposition 4.19 *Thompson's group F is non-amenable if and only if there exists a finite set H which can be decomposed as a disjoint union of sets H_1 and H_2 with $\sum_{g \in H_1} \pi(g) = \sum_{h \in H_2} \pi(h)$ and such that the closed ideal generated by $\sum_{g \in H_1} \lambda(g) - \sum_{h \in H_2} \lambda(h)$ coincides with $C^*_\lambda(T)$.*

Proof One part of the theorem follows from the draft [12] of Haagerup and Olesen (see also [20, Section 4.2]).

It is left to show that if F is not amenable then there is a finite set H which satisfies the conditions of the theorem. Let x_1 and x_2 be the standard generators of the copy of Thompson's group F in T, supported in $[\frac{1}{2}, 1] \subset S^1$, with supports in $[\frac{1}{2}, 1]$ and $[\frac{3}{4}, 1]$ respectively. Let $g_1 = x_1^r$ and $g_2 = x_2^r$ be the conjugated copies of these generators, where r is rotation by $1/2$, so that g_1 and g_2 act on the interval $[0, \frac{1}{2}] \subset S^1$ and generate a copy of F there, with trivial action on $[\frac{1}{2}, 1]$. Define $H = \{2e, g_1, g_2, x_1, x_2, g_1x_1, g_2x_2\}$. Define $H_1 = \{g_1, g_2, x_1, x_2\}$ and H_2 to be the rest of the set H. Obviously, $\sum_{g \in H_1} \pi(g) - \sum_{h \in H_2} \pi(h) = 0$.

Let us show that the ideal, J, generated by $\sum_{g \in H_1} \lambda(g) - \sum_{h \in H_2} \lambda(h)$ is the whole reduced C^*-algebra of T. Note that $\|1 + \lambda(g_1) + \lambda(g_2)\| \leq C < 3$ by assumption. Moreover, the point $p = \frac{7}{8}$ is not fixed by any element from the set $E = \{x_1, x_2, g_1x_1, g_2x_2\}$. Thus we can apply Lemma

4.7 for the set E: let $\varepsilon > 0$ and let g and $c_1, \ldots, c_n \in C_T(g)$ be such that

$$\|\sum_{s\in E}\sum_{i=1}^{n} \lambda((sg)^{c_i})\| \leq \varepsilon n.$$

Note that the element $b = \lambda(g - \frac{1}{3}[e + g_1 + g_2]g + \frac{1}{3}\sum_{s\in E} sg)$ is in J, thus $\frac{1}{n}\sum_{i=1}^{n} \lambda(c_i)b\lambda({c_i}^{-1}) \in J$. The distance between the element $\frac{1}{n}\sum_{i=1}^{n} \lambda(c_i)b\lambda({c_i}^{-1})$ and $\lambda(g)$ is strictly smaller than 1 for large n. Indeed,

$$\|\lambda(g) - b\| \leq \frac{1}{3}\|1 + \lambda(g_1) + \lambda(g_2)\| + \frac{1}{3n}\|\sum_{s\in E}\sum_{i=1}^{n} \lambda((sg)^{c_i})\| \leq C + \varepsilon,$$

thus we have found an invertible element in J, therefore $J = C^*_\lambda(T)$. □

4.6 Some Questions

The last result in particular shows that our Condition 4.5 is almost (in some sense) enough to show that $C^*_\lambda(T)$ is simple. But not quite! This, together with our partial converse of the Haagerup-Olesen result, encourages us to ask the foliowing question.

Question 1 *Is the non-amenability of Thompson's group F equivalent to the simplicity of the algebra $C^*_\lambda(T)$.*

As the setup of Condition 4.5 is more flexible than that of Condition 4.2, we find Condition 4.5 easier to use. Still, we have not actually proven that the conditions are not equivalent. Thus it would be quite useful to give a positive answer to the following question.

Question 2 *Is there a group G which fails to satisfy Condition 4.2 but which does satisfy Condition 4.5?*

In [14], [13], and [15] several approaches to amenability via group actions were developed. It is an interesting question to relate these approaches to properties of a group's C^*-algebra.

Acknowledgments

The authors are grateful to Uffe Haagerup and Kristian Knudsen Olesen for sharing the early drafts of their work [12] with us, and to Martin

Kassabov and Justin Tatch Moore for feedback on this chapter while it was under construction.

References

[1] Bleak, C. *An exploration of normalish subgroups of R. Thompson's groups F and T*. Preprint available at `http://arxiv.org/abs/1603.01726`.

[2] Bleak, Collin, Kassabov, Martin and Matucci, Francesco. 2011. Structure theorems for groups of homeomorphisms of the circle. *Internat. J. Algebra Comput.*, **21**(6), 1007–1036.

[3] Breuillard, E., Kalantar, M., Kennedy, M. and Ozawa, N. 2017. C^*-simplicity and the unique trace property for discrete groups. *Publ. Math. Inst. Hautes Études Sci.*, **126**, 35–71.

[4] Brin, Matthew G. 2005. Coherence of associativity in categories with multiplication. *J. Pure Appl. Algebra*, **198**(1-3), 57–65.

[5] Brin, Matthew G. and Bowlin, Garry. 2013. Coloring planar graphs via colored paths in the associahedra. *Internat. J. Algebra Comput.*, **23**(6), 1337–1418.

[6] Brin, Matthew G. and Squier, Craig C. 1985. Groups of piecewise linear homeomorphisms of the real line. *Invent. Math.*, **79**(3), 485–498.

[7] Cannon, J. W., Floyd, W. J. and Parry, W. R. 1996. Introductory notes on Richard Thompson's groups. *Enseign. Math. (2)*, **42**(3-4), 215–256.

[8] Dehornoy, Patrick. 2005. Geometric presentations for Thompson's groups. *J. Pure Appl. Algebra*, **203**(1-3), 1–44.

[9] Fricke, Robert and Klein, Felix. 1965. *Vorlesungen über die Theorie der automorphen Funktionen. Band 1: Die gruppentheoretischen Grundlagen. Band II: Die funktionentheoretischen Ausführungen und die Andwendungen.* Bibliotheca Mathematica Teubneriana, Bände 3, vol. 4. Johnson Reprint Corp.; B. G. Teubner Verlagsgesellschaft.

[10] Geoghegan, Ross and Guzmán, Fernando. 2006. Associativity and Thompson's group. Pages 113–135 of: *Topological and asymptotic aspects of group theory.* Contemp. Math., vol. 394. Amer. Math. Soc.

[11] Ghys, Étienne and Sergiescu, Vlad. 1987. Sur un groupe remarquable de difféomorphismes du cercle. *Comment. Math. Helv.*, **62**(2), 185–239.

[12] Haagerup, Uffe and Olesen, Kristian Knudsen. *On conditions towards the non-amenability of Richard Thompson's group F*. In preparation.

[13] Juschenko, Kate and de la Salle, Mikael. 2015. Invariant means for the wobbling group. *Bull. Belg. Math. Soc. Simon Stevin*, **22**(2), 281–290.

[14] Juschenko, Kate and Monod, Nicolas. 2013. Cantor systems, piecewise translations and simple amenable groups. *Ann. of Math. (2)*, **178**(2), 775–787.

[15] Juschenko, Kate, Nekrashevych, Volodymyr and de la Salle, Mikael. 2016. Extensions of amenable groups by recurrent groupoids. *Invent. Math.*, **206**(3), 837–867.

[16] Kalantar, Mehrdad and Kennedy, Matthew. 2017. Boundaries of reduced C^*-algebras of discrete groups. *J. Reine Angew. Math.*, **727**, 247–267.

[17] Kennedy, M. *An intrinsic characterisation of C^*-simplicity.* Preprint available at `http://arxiv.org/abs/1509.01870v4`.

[18] Kesten, Harry. 1959. Symmetric random walks on groups. *Trans. Amer. Math. Soc.*, **92**, 336–354.

[19] Le Boudec, A. 2017. C^*-simplicity and the amenable radical. *Invent. Math.*, **209**(1), 159–174.

[20] Olesen, Kristian Knudsen. 2016. *The Thompson groups.* Ph.D. thesis, University of Copenhagen.

[21] Powers, Robert T. 1975. Simplicity of the C^*-algebra associated with the free group on two generators. *Duke Math. J.*, **42**, 151–156.

5

Local Similarity Groups with Context-free Co-word Problem

Daniel Farley

Abstract

Let G be a group, and let S be a finite subset of G that generates G as a monoid. The *co-word problem* is the collection of words in the free monoid S^* that represent non-trivial elements of G.

A current conjecture, based originally on a conjecture of Lehnert and modified into its current form by Bleak, Matucci, and Neunhöffer, says that Thompson's group V is a universal group with context-free co-word problem. It is thus conjectured that a group has a context-free co-word problem exactly if it is a finitely generated subgroup of V.

Hughes introduced the class $\mathcal{FSS}$ of groups that are determined by finite similarity structures. An $\mathcal{FSS}$ group acts by local similarities on a compact ultrametric space. Thompson's group V is a representative example, but there are many others.

We show that $\mathcal{FSS}$ groups have context-free co-word problem under a minimal additional hypothesis. As a result, we can specify a subfamily of $\mathcal{FSS}$ groups that are potential counterexamples to the conjecture.

5.1 Introduction

Let G be a group, and let S be a finite subset that generates G as a monoid. The *word problem of G with respect to S*, denoted $\mathrm{WP}_S(G)$, is the collection of all positive words w in S such that w represents the identity in G; the *co-word problem of G with respect to S*, denoted $\mathrm{CoWP}_S(G)$, is the set of all positive words that represent non-trivial elements of G. In this point of view, both the word and the co-word problem of G are formal languages, which suggests the question of placing these problems within the Chomsky hierarchy of languages (as described, for instance, in [6]). The locations of the formal languages $\mathrm{WP}_S(G)$ and $\mathrm{CoWP}_S(G)$ within the hierarchy do not depend on the choice of finite generating set S.

Anisimov [1] proved that $\mathrm{WP}_S(G)$ is a regular language if and only if G is finite. A celebrated theorem of Muller and Schupp [14] says that a finitely generated group G has context-free word problem if and only if it is virtually free. (A language is *context-free* if it is recognized by a pushdown automaton.) In this case, as noted in [9], the word problem is actually a deterministic context-free language. Shapiro [18] described sufficient conditions for a group to have a context-sensitive word problem.

Since the classes of regular, deterministic context-free, and context-sensitive languages are all closed under taking complements, it is of no additional interest to study groups with regular, deterministic context-free, or context-sensitive co-word problems, since the classes of groups in question do not change. The (non-deterministic) context-free languages are not closed under taking complements, however, so groups with context-free co-word problem are not (a priori, at least) the same as groups with context-free word problem.

Holt, Rees, Röver, and Thomas [9] introduced the class of groups with context-free co-word problem, denoted co$\mathcal{CF}$. They proved that all finitely generated virtually free groups are co$\mathcal{CF}$, and that the class of co$\mathcal{CF}$ groups is closed under taking finite direct products, passage to finitely generated subgroups, passage to finite index overgroups, and taking restricted wreath products with virtually free top group. They proved some negative results as well: for instance, the Baumslag–Solitar groups $BS(m,n)$ are not co$\mathcal{CF}$ if $|m| \neq |n|$, and polycyclic groups are not co$\mathcal{CF}$ unless they are virtually abelian. They conjectured that co$\mathcal{CF}$ groups are not closed under the operation of taking free products, and indeed specifically conjectured that $\mathbb{Z} * \mathbb{Z}^2$ is not a co$\mathcal{CF}$ group.

Lehnert and Schweitzer [12] later showed that the Thompson group V (as described in [7]) is co$\mathcal{CF}$. Since V seems to contain many types of subgroup (among them all finite groups, all countable free groups, and all countably generated free abelian groups), this raised the possibility of showing that $\mathbb{Z}*\mathbb{Z}^2$ is co$\mathcal{CF}$ by embedding the latter group into V. Bleak and Salazar-Díaz [4], motivated at least in part by these considerations, proved that $\mathbb{Z} * \mathbb{Z}^2$ does not embed in V (leaving the conjecture from [9] open), and also established the existence of many embeddings into V. The basic effect of their embedding theorems was to show that the class $\mathcal{V}$ of finitely generated subgroups of Thompson's group V is closed under almost all of the same operations as those from [9], as listed above. More recently, Bennett and Bleak [2] have proved that $\mathcal{V}$ is also closed under restricted wreath products with virtually free top group (since virtually free groups are "demonstrative"– see [4] for definitions). It is thus now known that $\mathcal{V}$ and co$\mathcal{CF}$ are both closed under the group-theoretic constructions listed in [9] (and described above).

Conjecture 5.1 [11, 5] The classes $\mathcal{V}$ and co$\mathcal{CF}$ are the same; i.e., Thompson's group V is a universal co$\mathcal{CF}$ group.

Lehnert had conjectured in his thesis [11] that a certain closely related group Q of quasi-automorphisms of the infinite binary tree is a universal co$\mathcal{CF}$ group. Bleak, Matucci, and Neunhöffer [5] established the existence of embeddings from Q to V and from V to Q. As a result, Lehnert's conjecture is equivalent to Conjecture 5.1. We refer the reader to the excellent introductions of [5] and [4] for a more extensive discussion of these and related questions.

Here we show that many groups defined by finite similarity structures are contained in co$\mathcal{CF}$. The precise statement is as follows.

Main Theorem *Let X be a compact ultrametric space endowed with a finite similarity structure* Sim_X. *Assume that there are only finitely many* Sim_X*-classes of balls.*

For any finitely generated subgroup G of $\Gamma(\mathrm{Sim}_X)$ *and finite subset S of G that generates G as a monoid, the co-word problem* $\mathrm{CoWP}_S(G) = \{w \in S^* \mid w \neq 1_G\}$ *is a context-free language.*

The groups defined by finite similarity structures (or $\mathcal{FSS}$ groups) were first studied by Hughes [10], who showed that all $\mathcal{FSS}$ groups act properly on CAT(0) cubical complexes and (therefore) have the Haagerup property. Farley and Hughes [8] proved that a class of $\mathcal{FSS}$ groups have type $\mathcal{F}_\infty$. All of the latter groups satisfy the hypotheses of the Main Theorem, so all are also co$\mathcal{CF}$ groups. (We note that the Main Theorem also covers V as a special case.)

The class of $\mathcal{FSS}$ groups is not well understood, but we can specify a certain subclass that shows promise as a source of counterexamples to Conjecture 5.1. These are the Nekrashevych–Röver examples from [8] and [10]. The results of [8] show that most of these examples are not isomorphic to V (nor to the n-ary versions of V), and it is not difficult to show that they do not contain V as a subgroup of finite index. It seems to be unknown whether there are any embeddings of these groups into V. Our Main Theorem therefore leaves Conjecture 5.1 open.

(The Nekrashevych-Röver examples considered in [8] and [10] are not as general as the classes of groups from [17] and [15]; the finiteness of the similarity structures proves to be a somewhat restrictive hypothesis.)

We note that, even apart from the examples from our Main Theorem, there are differences between the class of known members of $\mathcal{V}$ and the class of known members of co$\mathcal{CF}$. Berns-Zieve, Fry, Gillings, Hoganson, and Mathews [3] have used the cloning systems of Witzel and Zaremsky [19] to describe a class of co$\mathcal{CF}$ groups having no known embeddings into V.

The proof of the Main Theorem closely follows the work of Lehnert and Schweitzer [12]. We identify two main ingredients of their proof.

1 All of the groups satisfying the hypothesis of the Main Theorem admit *test partitions* (Definition 5.21). That is, there is a finite partition of the compact ultrametric space X into balls, such that every non-trivial word in the generators of G has a cyclic rotation that moves at least one of the balls off itself, and

2 for each ball B from the test partition, there is a "B-witness automaton", which is a pushdown automaton that can witness an element $g \in G$ moving part of B off itself.

The Main Theorem follows very easily from (1) and (2). The proofs that (1) and (2) hold are complicated somewhat by the generality of our assumptions, but are already implicit in [12]. Most of the work goes into building the witness automata. We describe a stack language $\mathcal{L}$ that the witness automata use to describe, store, and manipulate metric balls in X. One slight novelty (not present or necessary in [12]) is that the witness automata write functions from the similarity structure on their stacks and make partial computations using these functions.

We briefly describe the structure of the chapter. Section 5.2 contains a summary of the relevant background, including string rewriting systems, pushdown automata, $\mathcal{FSS}$ groups, and standing assumptions. Section 5.3 contains a proof that the groups G admit test partitions, as described above. Section 5.4 describes the stack language for the witness automata, and Section 5.5 gives the construction of the witness automata. Section 5.6 collects the ingredients of the previous sections into a proof of the Main Theorem.

5.2 Background

5.2.1 String Rewriting Systems

The description of witness automata in Sections 5.4 and 5.5 will require some ideas from the theory of rewrite systems. We review these ideas here.

Definition 5.2 A *rewrite system* is a directed graph Γ. We write $a \to b$ if a and b are vertices of Γ and there is a directed edge from a to b. We write $a \to^* b$ if there is a directed path from a to b. The rewrite system Γ is called *locally confluent* if whenever $a \to b$ and $a \to c$, there is some $d \in \Gamma^0$ such that $c \to^* d$ and $b \to^* d$. The rewrite system is *confluent* if whenever $a \to^* b$ and $a \to^* c$, there is some $d \in \Gamma^0$ such that $c \to^* d$ and $b \to^* d$. The rewrite system Γ is *terminating* if there is no infinite directed path in Γ. If a rewrite system is both terminating and confluent, then we say that it is *complete*. A vertex of Γ is called *reduced* if it is not the initial vertex of any directed edge in Γ.

Theorem 5.3 *[16] (Newman's Lemma) Every terminating, locally confluent rewrite system is complete.* □

Remark The relation $\to$ generates an equivalence relation on the vertices of Γ. It is not difficult to see that each equivalence class in this equivalence relation contains a unique reduced element in the event that Γ is complete.

Definition 5.4 Let Σ be a finite set, called an *alphabet*. Let $\mathcal{L}$ be a subset of the free monoid Σ^*. Let $\mathcal{R}$ be a collection of relations (or *rewriting rules*) of the form $w_1 \to w_2$, where $w_1, w_2 \in \Sigma^*$. (Thus, the w_i are positive words in the alphabet Σ, either of which may be empty. The w_i are not required to be in $\mathcal{L}$.)

We define a *string rewriting system* as follows: The vertices are words from $\mathcal{L}$. For $u, v \in \mathcal{L}$, there is a directed edge $u \to v$ whenever there are words u', u'' such that $u = u'w_1u''$ and $v = u'w_2u''$, for some $w_1 \to w_2 \in \mathcal{R}$.

5.2.2 Pushdown Automata

Definition 5.5 Let S and Σ be finite sets. The set S is the *input alphabet* and Σ is the *stack alphabet*. The stack alphabet contains a special symbol, #, called the *initial stack symbol*.

A *(generalized) pushdown automaton (or PDA) over S and Σ* [6] is a finite labeled directed graph Γ endowed with an *initial state* $v_0 \in \Gamma^0$ and a (possibly empty) collection of *terminal states* $T \subseteq \Gamma^0$. Each directed edge is labeled by a triple $(s, w', w'') \in (S \cup \{\epsilon\}) \times \Sigma^* \times \Sigma^*$, where ϵ denotes an empty string.

Each PDA accepts languages either by terminal state, or by empty stack, and this information must be specified as part of the automaton's definition. See Definition 5.7.

Definition 5.6 Let Γ be a pushdown automaton. We describe a class of directed paths in Γ, called the *valid paths*, by induction on length. The path of length 0 starting at the initial vertex $v_0 \in \Gamma^0$ is valid; its *stack value* is $\# \in \Sigma^*$. Let $e_1 \ldots e_n$ $(n \geq 0)$ be a valid path in Γ, where e_1 is the edge that is crossed first. Let e_{n+1} be an edge whose initial vertex is the terminal vertex of e_n; we suppose that the label of e_{n+1} is (s, w', w''). The path $e_1 e_2 \ldots e_n e_{n+1}$ is also valid, provided that the stack value of $e_1 \ldots e_n$ has w' as a prefix; that is, if the stack value of $e_1 \ldots e_n$ has the form $w'\hat{w} \in \Sigma^*$. The stack value of $e_1 \ldots e_{n+1}$ is then $w''\hat{w}$. We let $\mathrm{val}(p)$ denote the stack value of a valid path p.

The *label* of a valid path $e_1 \ldots e_n$ is $s_n \ldots s_1$, where s_i is the first coordinate of the label for e_i (an element of S, or the empty string). The label of a valid path p will be denoted $\ell(p)$.

Definition 5.7 Let Γ be a PDA. The *language* $\mathcal{L}_\Gamma$ *accepted by* Γ is either

1 $\mathcal{L}_\Gamma = \{w \in S^* \mid w = \ell(p)$ for some valid path p with $\mathrm{val}(p) = \epsilon\}$, if Γ accepts by empty stack, or
2 $\mathcal{L}_\Gamma = \{w \in S^* \mid w = \ell(p)$ for some valid path p whose terminal vertex is in $T\}$, if Γ accepts by terminal state.

Definition 5.8 A subset of the free monoid S^* is called a *(non-deterministic) context-free language* if it is $\mathcal{L}_\Gamma$, for some pushdown automaton Γ.

Remark The class of languages that are accepted by empty stack (in the above sense) is the same as the class of languages that are accepted by terminal state. That is, given an automaton Γ' that accepts a language $\mathcal{L}$ by empty stack, there is another automaton Γ'' that accepts $\mathcal{L}$ by terminal state (and conversely).

Remark All of the automata considered in this chapter will accept by empty stack.

The functioning of an automaton Γ can be described in plain language as follows. We begin with a word $s_n \dots s_1 \in S^*$ written on an input tape, and the word $\# \in \Sigma^*$ written on the memory tape (or stack). We imagine the stack as a sequence of boxes extending indefinitely to our left, all empty except for the rightmost one, which has # written in it. Our automaton reads the input tape from right to left. It can read and write on the stack only from the left (i.e., from the leftmost non-empty box). Beginning in the initial state $v_0 \in \Gamma_0$, it can follow any directed edge e it chooses, provided that it meets the proper prerequisites: if the label of e is (s, w', w''), then s must be the rightmost remaining symbol on the input tape, and the word $w' \in \Sigma^*$ must be a prefix of the word written on the stack. If these conditions are met, then it can cross the edge e into the next state, simultaneously erasing the letter s from the input tape, erasing w' from the left end of the stack, and then writing w'' on the left end of the stack. The original input word is accepted if the automaton can reach a state with nothing left on its input tape, and nothing on its stack (not even the symbol #).

We note that a label such as $(\epsilon, \epsilon, w'')$ describes an empty set of prerequisites. Such an arrow may always be crossed, without reading the input tape or the stack, no matter whether one or the other is empty.

5.2.3 Review of Ultrametric Spaces and Finite Similarity Structures

We now give a quick review (without proofs) of finite similarity structures on compact ultrametric spaces, as defined in [10]. Most of this subsection is taken from [8].

Definition 5.9 An *ultrametric space* is a metric space (X, d) such that

$$d(x, y) \leq \max\{d(x, z), d(z, y)\},$$

for all $x, y, z \in X$.

Lemma 5.10 *Let X be an ultrametric space.*

1. *Let $N_\epsilon(x)$ be an open metric ball in X. If $y \in N_\epsilon(x)$, then $N_\epsilon(x) = N_\epsilon(y)$.*
2. *If B_1 and B_2 are open metric balls in X, then either the balls are disjoint, or one is contained in the other.*
3. *If X is compact, then each open ball B is contained in at most finitely many distinct open balls of X, and these form an increasing sequence:*

$$B = B_1 \subsetneq B_2 \subsetneq \dots \subsetneq B_n = X.$$

4 *If X is compact and x is not an isolated point, then each open ball $N_\epsilon(x)$ is partitioned by its maximal proper open subballs, which are finite in number.*

□

Convention 5.11 Throughout this chapter, "ball" will always mean "open ball". We note, however, that there is really no distinction between open and closed balls in a compact ultrametric space: by Lemma 5.10(3), every open ball $N_\epsilon(x)$ is the same as a closed ball $\overline{N}_{\epsilon'}(x) = \{z \in X \mid d(x,z) \leq \epsilon'\}$, for some $\epsilon' \leq \epsilon$. In particular, every open ball in a compact ultrametric space X is compact.

Definition 5.12 Let $f : X \to Y$ be a function between metric spaces. We say that f is a *similarity* if there is a constant $C > 0$ such that $d_Y(f(x_1), f(x_2)) = C d_X(x_1, x_2)$, for all x_1 and x_2 in X.

Definition 5.13 A *finite similarity structure for* X is a function Sim_X that assigns to each ordered pair B_1, B_2 of balls in X a (possibly empty) set $\mathrm{Sim}_X(B_1, B_2)$ of surjective similarities $B_1 \to B_2$ such that whenever B_1, B_2, B_3 are balls in X, the following properties hold:

1 (Finiteness) $\mathrm{Sim}_X(B_1, B_2)$ is a finite set;
2 (Identities) $\mathrm{id}_{B_1} \in \mathrm{Sim}_X(B_1, B_1)$;
3 (Inverses) if $h \in \mathrm{Sim}_X(B_1, B_2)$, then $h^{-1} \in \mathrm{Sim}_X(B_2, B_1)$;
4 (Compositions) if $h_1 \in \mathrm{Sim}_X(B_1, B_2)$ and $h_2 \in \mathrm{Sim}_X(B_2, B_3)$, then $h_2 h_1 \in \mathrm{Sim}_X(B_1, B_3)$;
5 (Restrictions) if $h \in \mathrm{Sim}_X(B_1, B_2)$ and $B_3 \subseteq B_1$, then

$$h_{|B_3} \in \mathrm{Sim}_X(B_3, h(B_3)).$$

Definition 5.14 A homeomorphism $h\colon X \to X$ is *locally determined by* Sim_X provided that for every $x \in X$, there exists a ball B' in X such that $x \in B'$, $h(B')$ is a ball in X, and $h|B' \in \mathrm{Sim}(B', h(B'))$.

Definition 5.15 The *finite similarity structure (FSS) group* $\Gamma(\mathrm{Sim}_X)$ is the set of all homeomorphisms $h\colon X \to X$ such that h is locally determined by Sim_X.

Remark The fact that $\Gamma(\mathrm{Sim}_X)$ is a group under composition is due to Hughes [10].

Definition 5.16 ([10], Definition 3.6) If $\gamma \in \Gamma(\mathrm{Sim}_X)$, then we can choose a partition of X by balls B such that, for each B, $\gamma(B)$ is a ball and $\gamma_{|B} \in \mathrm{Sim}_X(B, \gamma(B))$. Each element of this partition is called a *region* for γ.

Example 5.17 Suppose that X is the set of all infinite strings in the symbols 0 and 1. If w_1 and w_2 are two such infinite strings, then we set

$$d_X(w_1, w_2) = e^{-\ell},$$

where ℓ is the length of the longest prefix common to w_1 and w_2. The function $d_X : X \times X \to \mathbb{R}$ is an ultrametric on X. It is straightforward to verify that each metric ball in X takes the form $u* = \{w \in X \mid w = uv \text{ for some } v \in X\}$, for some finite string u of 0s and 1s.

We define a finite similarity structure Sim_X as follows. For each pair of balls $u*$, $v*$, let

$$\mathrm{Sim}_X(u*, v*) = \{\phi_{u,v}\},$$

where $\phi_{u,v}$ is the function which removes the prefix u from an infinite string $w \in u*$, and adds the prefix v in its place. The assignment Sim_X is indeed a finite similarity structure, and the group $\Gamma(\mathrm{Sim}_X)$ is identical to Thompson's group V in this case. If (T_1, T_2) is a labeled tree pair representing an element g of V, then the leaves of the domain tree T_1 correspond naturally to a choice of regions for g, where "region" is defined as above.

If we simply replace the set X in the above discussion by X_n, the set of all infinite strings in the symbols 0, ..., $n-1$, then we get the n-ary version V_n of Thompson's group V.

Let H be a group of permutations of the set $\{0, \ldots, n-1\}$. For finite strings u, v in the symbols 0, 1, ..., $n-1$ and $h \in H$, define

$$\mathrm{Sim}_{X_n}(u*, v*) = \{\phi_{u,v,h} \mid h \in H\},$$

where $\phi_{u,v,h}(ua_1a_2a_3\ldots) = vh(a_1)h(a_2)h(a_3)\ldots$ for $a_1a_2 \in X_n$. The assignment Sim_{X_n} is a finite similarity structure, and the associated group $\Gamma(\mathrm{Sim}_{X_n})$ is one of the Nekrashevych-Röver examples considered in [8] and [10]; these are the potential counterexamples to Conjecture 5.1 under consideration in this chapter.

5.2.4 Standing Assumptions

In this section, we set conventions that hold for the rest of the chapter.

Definition 5.18 We say that two balls B_1 and B_2 are in the same *Sim_X-class* if the set $\mathrm{Sim}_X(B_1, B_2)$ is non-empty.

Convention 5.19 We assume that X is a compact ultrametric space with finite similarity structure Sim_X. We assume that there are only finitely many Sim_X-classes of balls, represented by the balls

$$\tilde{B}_1, \ldots, \tilde{B}_k.$$

We let X be the chosen representative of its own similarity class, setting $X = \tilde{B}_1$. Let $[B]$ denote the Sim_X-class of a ball B.

Each ball $B \subseteq X$ is related to exactly one of the $\tilde{B}_i$. We choose (and fix) an element $f_B \in \mathrm{Sim}_X(\tilde{B}_i, B)$. We choose $f_{\tilde{B}_i} = id_{\tilde{B}_i}$ when $B = \tilde{B}_i$.

Each ball $\tilde{B}_i$ has a finite collection of ℓ_i maximal proper subballs, denoted

$$\tilde{B}_{i1}, \ldots, \tilde{B}_{i\ell_i}.$$

This numbering (of the balls $\tilde{B}_i$ and their maximal proper subballs) is fixed throughout the rest of the argument. (Note that, if $\tilde{B}_i$ is a singleton, then $\ell_i = 0$.) We let $\ell = \max\{\ell_1, \ldots, \ell_k\}$.

We will for the most part freely recycle the subscripts k and ℓ. However, for the reader's convenience, we note ahead of time that we will use k and ℓ with the above meaning in Definitions 5.28, 5.33, 5.34, and 5.38.

Convention 5.20 We will let G denote a finitely generated subgroup of $\Gamma(\mathrm{Sim}_X)$ (see Definition 5.15). We choose a finite set $S \subseteq G$ that generates G as a monoid, i.e., each element $g \in G$ can be expressed in the form $g = s_1 \ldots s_n$, where $s_i \in S$, $n \geq 0$, and only positive powers of the s_i are used. We choose (and fix) regions for each $s \in S$.

5.3 Test Partitions

Definition 5.21 Let $\mathcal{P}$ be a finite partition of X. We say that $\mathcal{P}$ is a *test partition* if, for any word $s_1 \ldots s_n$ in the generators S, whenever

$$s_i \ldots s_n s_1 \ldots s_{i-1}(\widehat{B}) = \widehat{B},$$

for all $i \in \{1, \ldots, n\}$ and $\widehat{B} \in \mathcal{P}$, then $s_1 \ldots s_n = 1_G$.

Lemma 5.22 *If X is a compact ultrametric space and $\epsilon > 0$, then $\{N_\epsilon(x) \mid x \in X\}$ is a finite partition of X by open balls.*

Proof This follows easily from Lemma 5.10(1). □

Definition 5.23 Let $\epsilon_1 > 0$ be chosen so that, for every $s \in S$ and $x \in X$, there exists a region R of s such that

$$N_{\epsilon_1}(x) \subseteq R.$$

We let $\mathcal{P}_{big} = \{N_{\epsilon_1}(x) \mid x \in X\}$. This is the *big partition.*

We note that such an ϵ_1 can always be chosen by, for instance, letting $\epsilon_1 \leq \min\{diam(R) \mid R$ is a region for some $s \in S\}$. (Here we must choose a positive diameter d for each singleton region $\{z\}$ in such a way that $N_d(z) = \{z\}$.)

Lemma 5.24 *Let B be a compact ultrametric space, and let Γ be a finite group of isometries of B. There is an $\epsilon > 0$ such that if $\gamma \in \Gamma$ acts trivially on $\{N_\epsilon(x) \mid x \in B\}$, then $\gamma = 1_\Gamma$.*

Proof For each non-trivial $\gamma \in \Gamma$, there is $x_\gamma \in X$ such that $\gamma(x_\gamma) \neq x_\gamma$. We choose $\epsilon_\gamma > 0$ satisfying

$$N_{\epsilon_\gamma}(x_\gamma) \cap N_{\epsilon_\gamma}(\gamma(x_\gamma)) = \emptyset.$$

We set $\epsilon = \min\{\epsilon_\gamma \mid \gamma \in \Gamma - \{1_\Gamma\}\}$. Now suppose that $\gamma \neq 1_\Gamma$ and γ acts trivially on $\{N_\epsilon(x) \mid x \in X\}$. Thus $\gamma(N_\epsilon(x_\gamma)) = N_\epsilon(x_\gamma)$, so $N_\epsilon(\gamma(x_\gamma)) \cap N_\epsilon(x_\gamma) \neq \emptyset$, but

$$N_\epsilon(\gamma(x_\gamma)) \cap N_\epsilon(x_\gamma) \subseteq N_{\epsilon_\gamma}(x_\gamma) \cap N_{\epsilon_\gamma}(\gamma(x_\gamma)) = \emptyset,$$

a contradiction. □

Definition 5.25 Write $\mathcal{P}_{big} = \{B_1, \ldots, B_\ell\}$. For each B_k $(1 \leq k \leq \ell)$, we can choose $\hat{\epsilon}_k \leq \epsilon_1$ to meet the conditions satisfied by ϵ in the previous lemma, for $\Gamma = \mathrm{Sim}_X(B_k, B_k)$. Let $\epsilon_2 = \min\{\hat{\epsilon}_1, \ldots, \hat{\epsilon}_\ell\}$. Let $\mathcal{P}_{small} = \{N_{\epsilon_2}(x) \mid x \in X\}$. This is the *small partition.*

Proposition 5.26 *The small partition $\mathcal{P}_{small}$ is a test partition.*

Proof Let $s_1 \ldots s_n$ be a word in the generators S; we assume $s_1 \ldots s_n \neq 1$. We suppose, for a contradiction, that for all $\widehat{B} \in \mathcal{P}_{small}$,

$$s_i \ldots s_n s_1 \ldots s_{i-1}(\widehat{B}) = \widehat{B},$$

for all $i \in \{1, \ldots, n\}$. Since $s_1 \ldots s_n \neq 1$, we can find $x \in X$ such that $s_1 \ldots s_n(x) \neq x$.

Sublemma 5.27 *Fix $s_1 \ldots s_n \in S^+$. For each $x \in X$, there is an open ball B, with $x \in B$, such that:*

1 $s_{i+1} \ldots s_n(B)$ lies in a region of s_i, for $i = 1, \ldots, n$, and

2 either $B \in \mathcal{P}_{big}$ or $s_j \dots s_n(B) \in \mathcal{P}_{big}$ for at least one $j \in \{1, \dots, n\}$.

Proof We first prove that, for any $x \in X$, there is a ball neighborhood B of x satisfying (1). We choose and fix $x \in X$.

Consider the elements $s_1, s_2, \dots, s_n \in G$. We first observe that there is a constant $C \geq 1$ such that if any ball B' lies inside a region for s_i (for any $i \in \{1, \dots, n\}$), then s_i stretches B' by a factor of no more than C. Next, observe that there is a constant D such that any ball of diameter less than or equal to D lies inside a region for s_i, for all $i \in \{1, \dots, n\}$. It follows easily that any ball of diameter less than D/C^{n-1} satisfies (1); we can clearly choose some such ball, B_1 to be a neighborhood of x. We note that if a ball satisfies (1), then so does every subball.

Let

$$B_1 \subsetneq B_2 \subsetneq B_3 \subsetneq \dots \subsetneq B_m = X$$

be the collection of all balls containing B_1. (Thus, each B_λ is a maximal proper subball inside $B_{\lambda+1}$, for $\lambda = 1, \dots, m-1$.) There is a largest $\alpha \in \{1, \dots, m\}$ such that B_α satisfies (1).

If $\alpha = m$, then the entire composition $s_1 \dots s_n \in \mathrm{Sim}_X(X, X)$. We then take $\widetilde{B} \in \mathcal{P}_{big}$ such that $s_1 \dots s_n(x) \in \widetilde{B}$. The required B is $(s_1 \dots s_n)^{-1}(\widetilde{B})$.

Now assume that $\alpha < m$. There is some $j \in \{1, \dots, n\}$ such that $s_{j+1} \dots s_n(B_{\alpha+1})$ is a ball and

$$(s_{j+1} \dots s_n)_{|B_{\alpha+1}} \in \mathrm{Sim}_X(B_{\alpha+1}, s_{j+1} \dots s_n(B_{\alpha+1})),$$

but $(s_{j+1} \dots s_n)(B_{\alpha+1})$ properly contains a region for s_j. (Note that $s_{n+1} \dots s_n(B_{\alpha+1}) = B_{\alpha+1}$, by our conventions.) Let $R_1, \dots, R_\beta$ be the regions of s_j that are contained in $(s_{j+1} \dots s_n)(B_{\alpha+1})$. We must have $R_\delta \subseteq (s_{j+1} \dots s_n)(B_\alpha)$ for some δ (by maximality of $(s_{j+1} \dots s_n)(B_\alpha)$ in $(s_{j+1} \dots s_n)(B_{\alpha+1})$); the reverse containment $(s_{j+1} \dots s_n)(B_\alpha) \subseteq R_\delta$ follows, since $(s_{j+1} \dots s_n)(B_\alpha)$ is contained in a region for s_j by our assumptions.

Now note that R_δ is partitioned by elements of $\mathcal{P}_{big}$; there is some $\widetilde{B} \subseteq R_\delta$ ($\widetilde{B} \in \mathcal{P}_{big}$) such that $s_{j+1} \dots s_n(x) \in \widetilde{B}$. We have that the map $s_{j+1} \dots s_n : B_\alpha \to R_\delta$ is a map from the similarity structure. The required ball B is $(s_{j+1} \dots s_n)^{-1}(\widetilde{B})$. □

Apply the sublemma to x: there is B (an open ball) with the given properties. Let $j \in \{1, \dots, n+1\}$ be such that

$$(s_j \dots s_n)_{|B} \in \mathrm{Sim}_X(B, s_j \dots s_n(B)),$$

where $s_j \ldots s_n(B) \in \mathcal{P}_{big}$ and the case $j = n+1$ corresponds to the case in which $s_j \ldots s_n$ is an empty product.

Since $s_j \ldots s_n(B)$ is invariant under every cyclic permutation of $s_1 \ldots s_n$ by our assumption,

$$s_j \ldots s_n s_1 \ldots s_{j-1}(s_j \ldots s_n(B)) = s_j \ldots s_n(B),$$

so $s_1 \ldots s_n(B) = B$.

Our assumptions imply that $(s_j \ldots s_n)_{|B} : B \to s_j \ldots s_n(B) = \widetilde{B}$ and $(s_1 \ldots s_{j-1})_{|\widetilde{B}} : \widetilde{B} \to B$ are both in Sim_X, and both are bijections.

Consider the element $(s_j \ldots s_n s_1 \ldots s_{j-1})_{|\widetilde{B}} \in \mathrm{Sim}_X(\widetilde{B}, \widetilde{B})$. It must be that $(s_j \ldots s_n s_1 \ldots s_{j-1})_{|\widetilde{B}} \neq 1_{\widetilde{B}}$; if $(s_j \ldots s_n s_1 \ldots s_{j-1})_{|\widetilde{B}} = 1_{\widetilde{B}}$, then

$$s_j \ldots s_n s_1 \ldots s_{j-1}(s_j \ldots s_n)(x) = s_j \ldots s_n(x),$$

which implies that $s_1 \ldots s_n(x) = x$, a contradiction.

Now, since $(s_j \ldots s_n s_1 \ldots s_{j-1})_{|\widetilde{B}} \neq 1_{\widetilde{B}}$, it moves some element of $\mathcal{P}_{small}$. □

5.4 A Language for Sim_X

In this section, we introduce languages $\mathcal{L}_{red}$ and $\mathcal{L}$. The language $\mathcal{L}$ will serve as the stack language for the witness automata of Section 5.5. The language $\mathcal{L}_{red}$ consists of the reduced elements of $\mathcal{L}$; it is useful because there is a one-to-one correspondence between elements of $\mathcal{L}_{red}$ and metric balls in X.

5.4.1 The Languages $\mathcal{L}_{red}$ and $\mathcal{L}$

We want the language $\mathcal{L}_{red}$ to give a recursively defined "address" to each metric ball B in X. Our approach is as follows. If $B = X$, then B will be denoted by $A_{1,\emptyset}$, since B is in the same Sim_X-class as the first ball (i.e., $X = \tilde{B}_1$, as in Convention 5.19), and B is not a proper subball of any other ball, as indicated by the subscript $\emptyset$. If B is instead a maximal proper subball of X (so $B = \tilde{B}_{1n}$, for some $n \in \{1, \ldots, \ell_1\}$, as in Convention 5.19) and B is in the same Sim_X-class as $\tilde{B}_i$, then B will be denoted by the string $A_{1,\emptyset}A_{i,n}$. Here the maximal proper prefix $A_{1,\emptyset}$ tells us the address of the minimal ball properly containing B (i.e., the address of X), and $A_{i,n}$ tells us both the Sim_X-class of B (i.e., $[B] = [\tilde{B}_i]$) and the number n of B as a subball of X (which was decided upon in Convention 5.19). The description of all remaining metric balls proceeds inductively. We now offer the details.

Definition 5.28 We define a language $\mathcal{L}_{red}$ as follows. The alphabet Σ for $\mathcal{L}_{red}$ consists of the symbols:

1 $\#$, the initial stack symbol;
2 $A_{1,\emptyset}$;
3 $A_{i,n}$, $i \in \{1, \ldots, k\}$, $n \in \{1, \ldots, \ell\}$.

(We refer the reader to Convention 5.19 for the meanings of k and ℓ.) The language $\mathcal{L}_{red}$ consists of all words of the form

$$A_{1,\emptyset} A_{i_1,n_1} A_{i_2,n_2} \ldots A_{i_m,n_m} \#,$$

where $m \geq 0$ and $[\tilde{B}_{i_{s-1}n_s}] = [\tilde{B}_{i_s}]$ for $s = 1, \ldots, m$. (Here, and in what follows, we make the convention that $i_0 = 1$; i.e., that $X = \tilde{B}_{i_0}$.)

The language $\mathcal{L}$ also uses symbols of the form $[f]$, where we have $f \in \mathrm{Sim}_X(\tilde{B}_i, \tilde{B}_i)$. The general element of $\mathcal{L}$ takes the form

$$A_{1,\emptyset} w_0 A_{i_1,n_1} w_1 A_{i_2,n_2} w_2 \ldots A_{i_{m-1},n_{m-1}} w_{m-1} A_{i_m,n_m} w_m \#,$$

where each w_j ($j \in \{0, 1, \ldots, m\}$) is a word in the symbols $\{[f] \mid f \in \mathrm{Sim}_X(\tilde{B}_{i_j}, \tilde{B}_{i_j})\}$, and some or all of the w_j might be empty.

Remark As noted before Definition 5.28, the letter $A_{i,n}$ signifies a ball of similarity class $[\tilde{B}_i]$; the n signifies that it is the nth maximal proper subball of the ball before it in the sequence. The letter $A_{1,\emptyset}$ signifies the top ball, X.

The condition $[\tilde{B}_{i_{s-1}n_s}] = [\tilde{B}_{i_s}]$ for $s = 1, \ldots, m$ is designed to ensure that each ball has the correct type; i.e., that the sequence encodes consistent information about the similarity types of subballs.

We note that the alphabet Σ potentially contains symbols that can never occur in the language $\mathcal{L}_{red}$. (For instance, the letter $A_{i,n}$ can never occur in a word of $\mathcal{L}_{red}$ if a ball in the Sim_X-class of $\tilde{B}_i$ never occurs as the nth maximal proper subball in any of the balls $\tilde{B}_1, \ldots, \tilde{B}_k$ listed in Convention 5.19.)

Also note that, if $\tilde{B}_i$ is a singleton, then any letter $A_{i,n}$ occurring in a word $w \in \mathcal{L}_{red}$ must occur immediately before the end of the stack symbol $\#$.

Definition 5.29 Let $\mathcal{B}_X$ denote the collection of all metric balls in X. We define an *evaluation map* $E : \mathcal{L} \to \mathcal{B}_X$ by sending

$$w = A_{1,\emptyset} w_0 A_{i_1,n_1} w_1 A_{i_2,n_2} w_2 \ldots A_{i_{m-1},n_{m-1}} w_{m-1} A_{i_m,n_m} w_m \#$$

to

$$E(w) = \left(f_{\tilde{B}_{1n_1}} \circ f_{w_0} \circ f_{\tilde{B}_{i_1 n_2}} \circ f_{w_1} \circ \ldots \circ f_{w_{m-1}} \circ f_{\tilde{B}_{i_{m-1} n_m}} \right) (\tilde{B}_{i_m}),$$

where $f_{w_i} = f_{j_1} \circ f_{j_2} \circ \ldots \circ f_{j_\alpha}$ if $w_i = [f_{j_1}][f_{j_2}] \ldots [f_{j_\alpha}]$.

Definition 5.30 Let $w, w' \in \mathcal{L}$. We say that w' is a *prefix* of w if w' with the initial stack symbol # omitted is a prefix of w in the usual sense; that is $w = w'u$, for some string $u \in \Sigma^*$.

Proposition 5.31 *The function $E : \mathcal{L}_{red} \to \mathcal{B}_X$ is a bijection. Moreover, a word $w' \in \mathcal{L}_{red}$ is a proper prefix of $w \in \mathcal{L}_{red}$ if and only if $E(w)$ is a proper subball of $E(w')$, and w' is a maximal proper prefix of w if and only if $E(w)$ is a maximal proper subball of $E(w')$.*

Proof We first prove surjectivity. Let B be a ball in X. We let

$$B = B_m \subsetneq B_{m-1} \subsetneq B_{m-2} \subsetneq \ldots \subsetneq B_0 = X$$

be the collection of all balls in X that contain B_m. (Thus, B_i is a maximal proper subball in B_{i-1} for $i = 1, \ldots, m$.)

In the diagram

$$\begin{array}{ccccccc}
\tilde{B}_{i_m} & & \tilde{B}_{i_{m-1}} & & \tilde{B}_{i_{m-2}} & \cdots & \tilde{B}_{i_0} \\
\downarrow f_{B_m} & & \downarrow f_{B_{m-1}} & & \downarrow f_{B_{m-2}} & & \downarrow f_X \\
B_m & \longrightarrow & B_{m-1} & \longrightarrow & B_{m-2} & \longrightarrow \cdots \longrightarrow & X,
\end{array}$$

the balls $\tilde{B}_{i_j}$ and the maps f_{B_j} are the ones given in Convention 5.19; the unlabeled arrows are inclusions. Note, in particular, that the maps f_{B_j} are bijections taken from the Sim_X-structure, and that f_X is the identity map. If we follow the arrows from $\tilde{B}_{i_j}$ to $\tilde{B}_{i_{j-1}}$, the corresponding composition is a member of Sim_X that carries the ball $\tilde{B}_{i_j}$ to a maximal proper subball of $\tilde{B}_{i_{j-1}}$. Supposing that the number of the latter maximal proper subball is n_j (see Convention 5.19), we obtain a diagram

$$\begin{array}{ccccccc}
\tilde{B}_{i_m} & \xrightarrow{I_m} & \tilde{B}_{i_{m-1}} & \xrightarrow{I_{m-1}} & \tilde{B}_{i_{m-2}} & \xrightarrow{I_{m-2}} \cdots \xrightarrow{I_1} & \tilde{B}_{i_0} \\
\downarrow f_{B_m} & & \downarrow f_{B_{m-1}} & & \downarrow f_{B_{m-2}} & & \downarrow f_X \\
B_m & \longrightarrow & B_{m-1} & \longrightarrow & B_{m-2} & \longrightarrow \cdots \longrightarrow & X,
\end{array}$$

where $I_j = f_{\tilde{B}_{i_{j-1} n_j}}$, for $j = 1, \ldots, m$. This diagram commutes "up to images": that is, if we start at a given node in the diagram, then the image of that first node in any other node is independent of path.

(The diagram is not guaranteed to commute in the usual sense.) Set $w = A_{1,\emptyset}A_{i_1,n_1}\ldots A_{i_m,n_m}$. We note that

$$\begin{aligned}E(w) &= (I_1 \circ I_2 \circ \ldots \circ I_m)(\tilde{B}_{i_m}) \\ &= (f_X \circ I_1 \circ I_2 \circ \ldots \circ I_m)(\tilde{B}_{i_m}) \\ &= f_{B_m}(\tilde{B}_{i_m}) \\ &= B_m,\end{aligned}$$

where the first equality is the definition of $E(w)$, the second follows since $f_X = \mathrm{id}_X$, the third follows from the commutativity of the diagram up to images, and the fourth follows from surjectivity of f_{B_m}. This proves that $E : \mathcal{L}_{red} \to \mathcal{B}_X$ is surjective.

Before proving injectivity of E, we note that, for a given

$$w = A_{1,\emptyset}A_{i_1,n_1}\ldots A_{i_m,n_m}$$

and associated

$$E(w) = \left(f_{\tilde{B}_{i_0 n_1}} \circ f_{\tilde{B}_{i_1 n_2}} \circ \ldots \circ f_{\tilde{B}_{i_{m-1} n_m}}\right)(\tilde{B}_{i_m}),$$

each of the functions I_s (which, by definition, is the same as $f_{\tilde{B}_{i_{s-1} n_s}} : \tilde{B}_{i_s} \to \tilde{B}_{i_{s-1} n_s}$, but with a different codomain) maps its domain onto a maximal proper subball of its codomain. As a result, a word w' is a proper prefix of w if and only if $E(w)$ is a proper subball of $E(w')$, and w' is a maximal proper prefix of w if and only if $E(w)$ is a maximal proper subball of $E(w')$.

Suppose now that $E(w_1) = E(w_2)$, for some $w_1, w_2 \in \mathcal{L}_{red}$, $w_1 \neq w_2$. By the above discussion, we can assume that neither w_1 nor w_2 is a prefix of the other. Let w_3 be the largest common prefix of w_1 and w_2. Let $E(w_3) = B$. Since $w_1 = w_3w'$ and $w_2 = w_3w''$ for non-trivial strings w' and w'' with different initial symbols, $E(w_1)$ and $E(w_2)$ are disjoint proper subballs of $E(w_3)$. □

Definition 5.32 Let B be a ball in X. The *address* of B is the inverse image of B under the evaluation map $E : \mathcal{L}_{red} \to \mathcal{B}_X$, but with the initial stack symbol omitted. We write *addr*(B).

5.4.2 A String Rewriting System Based on $\mathcal{L}$

In this subsection, we describe a string rewriting system with underlying vertex set $\mathcal{L}$. The witness automata of Section 5.5 will use this rewrite system to perform partial calculations in Sim_X on their stacks.

Definition 5.33 Define

$$[\cdot] : \bigcup_{(B_1,B_2)} \mathrm{Sim}_X(B_1, B_2) \to \{[f] \in \Sigma \mid f \in \mathrm{Sim}_X(\tilde{B}_j, \tilde{B}_j), j \in \{1, \ldots, k\}\}$$

by the rule $[h] = [f_{B_2}^{-1} h f_{B_1}]$, for $h \in \mathrm{Sim}_X(B_1, B_2)$. (We recall that k is the number of Sim_X-classes of balls in X; see Convention 5.19.) The union is over all pairs of balls $B_1, B_2 \subseteq X$.

If $[h] = [f]$, where $f \in \mathrm{Sim}_X(\tilde{B}_j, \tilde{B}_j)$ for some $j \in \{1, \ldots, k\}$, then f is the *standard representative* of h, and $[f]$ is the *standard form* for $[h]$.

Remark If $f \in \mathrm{Sim}_X(\tilde{B}_j, \tilde{B}_j)$ for some $j \in \{1, \ldots, k\}$, then we sometimes confuse $[f]$ with f itself; this is justified by our choices in Convention 5.19.

Definition 5.34 Define a string rewriting system $(\mathcal{L}, \to)$ as follows. The vertices are elements of the language $\mathcal{L}$. There are four families of rewriting rules:

1 (Restriction)

$$[f]A_{i_s,n_s} \to A_{i_s,f(n_s)}[f_{|\tilde{B}_{i_{s-1}} n_s}],$$

where $[f]$ is a standard form; i.e., $f \in \mathrm{Sim}_X(\tilde{B}_{i_{s-1}}, \tilde{B}_{i_{s-1}})$;

2 (Group multiplication)

$$[f_1][f_2] \to [f_1 \circ f_2],$$

where $f_1, f_2 \in \mathrm{Sim}_X(\tilde{B}_j, \tilde{B}_j)$, for some $j \in \{1, \ldots, k\}$;

3 (Absorption)

$$[f]\# \to \#,$$

for arbitrary $[f]$;

4 (Identities)

$$[id_{\tilde{B}_j}] \to \emptyset,$$

for $j = 1, \ldots, k$.

Remark We note that the total number of the above rules is finite, since there are only finitely many Sim_X-classes of balls.

Proposition 5.35 *The string rewriting system $(\mathcal{L}, \to)$ is locally confluent and terminating. Each reduced element of $\mathcal{L}$ is in $\mathcal{L}_{red}$. The function E is constant on equivalence classes modulo $\to$.*

Proof It is clear that each reduced element of $\mathcal{L}$ is in $\mathcal{L}_{red}$, and that $(\mathcal{L}, \to)$ is terminating. Local confluence of $(\mathcal{L}, \to)$ is clear, except for one case, which we will now consider.

Suppose that $w \in \mathcal{L}$ contains a substring of the form $[f][g]A_{i_s n_s}$. We can apply two different overlapping rewrite rules to w, one sending $[f][g]A_{i_s n_s}$ to $[f \circ g]A_{i_s n_s}$, and the other sending $[f][g]A_{i_s n_s}$ to $[f]A_{i_s g(n_s)}[g_{|\tilde{B}_{i_{s-1} n_s}}]$. We now need to show that both $[f \circ g]A_{i_s n_s}$ and $[f]A_{i_s g(n_s)}[g_{|\tilde{B}_{i_{s-1} n_s}}]$ flow to a common string. Note that

$$[f \circ g]A_{i_s n_s} \to A_{i_s f(g(n_s))}[(f \circ g)_{|\tilde{B}_{i_{s-1} n_s}}]$$

and

$$[f]A_{i_s g(n_s)}[g_{|\tilde{B}_{i_{s-1} n_s}}] \to A_{i_s f(g(n_s))}[f_{|\tilde{B}_{i_{s-1} g(n_s)}}][g_{|\tilde{B}_{i_{s-1} n_s}}].$$

It therefore suffices to demonstrate that the maps $[(f \circ g)_{|\tilde{B}_{i_{s-1} n_s}}]$ and $[f_{|\tilde{B}_{i_{s-1} g(n_s)}}][g_{|\tilde{B}_{i_{s-1} n_s}}]$ are equal. But this follows from the commutativity of the following diagram:

$$\begin{array}{ccccc}
\tilde{B}_{i_{s-1}} & \xrightarrow{[g_|]} & \tilde{B}_{i_{s-1}} & \xrightarrow{[f_|]} & \tilde{B}_{i_{s-1}} \\
\downarrow & & \downarrow & & \downarrow \\
\tilde{B}_{i_{s-1} n_s} & \xrightarrow{g} & g(\tilde{B}_{i_{s-1} n_s}) & \xrightarrow{f} & (f \circ g)(\tilde{B}_{i_{s-1} n_s}),
\end{array}$$

where the vertical arrows are the canonical identifications from Convention 5.19 (e.g., the first vertical arrow is $f_{\tilde{B}_{i_{s-1} n_s}}$). It now follows that $(\mathcal{L}, \to)$ is locally confluent and terminating.

We now prove that E is constant on the equivalence classes modulo $\to$. It is clear that applications of rules (2)-(4) do not change the value of E; we check that (1) also does not change the value of E. Suppose we are given

$$\tilde{B}_{i_m} \xrightarrow{I_m} \tilde{B}_{i_{m-1}} \xrightarrow{I_{m-1}} \ldots \longrightarrow \tilde{B}_{i_1} \xrightarrow{I_1} \tilde{B}_{i_0} = X ,$$

where each $I_j = f_{w_{m-1}} \circ f_{\tilde{B}_{i_{m-1} n_m}}$, and $(I_1 \circ \ldots \circ I_m)(\tilde{B}_{i_m})$ is therefore $E(w)$, for $w \in \mathcal{L}$ in the form given in Definition 5.28. We pick a particular $I_\alpha = f_{w_{\alpha-1}} \circ f_{\tilde{B}_{i_{\alpha-1} n_\alpha}}$, for some $\alpha \in \{1, \ldots, m\}$. We note that the map I_α corresponds to the substring $w_{\alpha-1} A_{i_\alpha n_\alpha}$ of w. We may assume that $w_{\alpha-1}$ has length 1 (after applying rewriting rules of the form (2)); we write f in place of $f_{w_{\alpha-1}}$, where $f \in \mathrm{Sim}_X(\tilde{B}_{i_{\alpha-1}}, \tilde{B}_{i_{\alpha-1}})$. The result of applying a rewrite rule of type (1) to $w_{\alpha-1}A_{i_\alpha n_\alpha}$ is the string $A_{i_\alpha f(n_\alpha)}[f_{|\tilde{B}_{i_{\alpha-1}n_\alpha}}]$. The latter string corresponds to the map

$$I'_\alpha = f_{\tilde{B}_{i_{\alpha-1}f(n_\alpha)}} \circ [f_{|\tilde{B}_{i_{\alpha-1}}n_\alpha}],$$

so we must show that $I'_\alpha = I_\alpha$. We consider the commutative diagram

$$\begin{array}{ccccc}
\tilde{B}_{i_\alpha} & \longrightarrow & \tilde{B}_{i_{\alpha-1}n_\alpha} & \longrightarrow & \tilde{B}_{i_{\alpha-1}} \\
\Big\downarrow{\scriptstyle [f_|]} & & \Big\downarrow{\scriptstyle f_{|\tilde{B}_{i_{\alpha-1}n_\alpha}}} & & \Big\downarrow{\scriptstyle f} \\
\tilde{B}_{i_\alpha} & \longrightarrow & \tilde{B}_{i_{\alpha-1}f(n_\alpha)} & \longrightarrow & \tilde{B}_{i_{\alpha-1}},
\end{array}$$

where the leftmost horizontal arrows are the canonical identifications of Convention 5.19 and the rightmost horizontal arrows are inclusions. If we follow the arrows in this rectangle from the upper left corner, down the left side, and across the bottom, the resulting map is I'_α; if we follow the arrows along the top and right side, the resulting map is I_α. This proves that $I'_\alpha = I_\alpha$, as required. □

5.4.3 The Action of Sim_X on $\mathcal{L}$

Definition 5.36 Let $f \in \mathrm{Sim}_X(B', B'')$, where B', B'' are arbitrary balls in X. Suppose that $addr(B') = \hat{w}$ and $addr(B'') = \tilde{w}$, where

$$\hat{w} = A_{1,\emptyset}A_{i_1n_1}\dots A_{i_mn_m} \quad \text{and} \quad \tilde{w} = A_{1,\emptyset}A_{j_1\ell_1}\dots A_{j_t\ell_t}.$$

Let $w \in \mathcal{L}$. For a word $w \in \mathcal{L}$, define a partial function $\phi_f : \mathcal{L} \to \mathcal{L}$ by the rule

$$\phi_f(w) = \tilde{w}[f]w'\#$$

if w has $\hat{w}$ as a prefix, i.e., if $w = \hat{w}w'\#$ for some string w'. Otherwise, $\phi_f(w)$ is undefined.

Proposition 5.37 *The expression $\phi_f(w)$ is defined if and only if $addr(B')$ is a prefix of $E(w)$. If $\phi_f(w)$ is defined, then*

$$E(\phi_f(w)) = f(E(w)).$$

Proof The first statement is straightforward.

Assume first that $w \in \mathcal{L}_{red}$. We have that

$$w = A_{1,\emptyset}A_{i_1n_1}\dots A_{i_mn_m}A_{i_{m+1}n_{m+1}}\dots A_{i_un_u}\#.$$

If we write I_v in place of $f_{\tilde{B}_{i_{v-1}n_v}}$ for $v \in \{1, \ldots, u\}$, then $E(w) = (I_1 \circ \ldots \circ I_u)(\tilde{B}_{i_u})$. If

$$B_u \subseteq B_{u-1} \subseteq \ldots \subseteq B_1 \subseteq X$$

is the sequence of all balls containing $B_u = E(w)$ (so that each ball is necessarily a maximal proper subball in the next), we have

$$\begin{array}{ccccccccc}
\tilde{B}_{i_u} & \xrightarrow{I_u} & \tilde{B}_{i_{u-1}} & \xrightarrow{I_{u-1}} & \tilde{B}_{i_{u-2}} & \xrightarrow{I_{u-2}} & \ldots & \xrightarrow{I_{m+1}} & \tilde{B}_{i_m} \\
\downarrow & & \downarrow & & \downarrow & & & & \downarrow \\
B_u & \longrightarrow & B_{u-1} & \longrightarrow & B_{u-2} & \longrightarrow & \ldots & \longrightarrow & B',
\end{array}$$

where the bottom horizontal arrows are inclusions, the vertical arrows are the canonical maps, and the diagram commutes up to images. Concatenating diagrams, we have

$$\begin{array}{ccccccccccc}
\tilde{B}_{i_u} & \xrightarrow{I_u} & \ldots & \xrightarrow{I_{m+1}} & \tilde{B}_{i_m} & \xrightarrow{[f]} & \tilde{B}_{j_t} & \xrightarrow{I'_t} & \ldots & \xrightarrow{I'_1} & X \\
\downarrow & & & & \downarrow & & \downarrow & & & & \downarrow {\scriptstyle f_X = \mathrm{id}_X} \\
B_u & \longrightarrow & \ldots & \longrightarrow & B' & \xrightarrow{f} & B'' & \longrightarrow & \ldots & \longrightarrow & X,
\end{array}$$

where the left rectangle is the previous diagram, the middle square commutes, and the right rectangle defines $E(\tilde{w})$ (i.e, the bottom horizontal arrows are inclusions, the vertical arrows are the canonical identifications, and the maps I'_β ($\beta \in \{1, \ldots, t\}$) are the ones from the definition of $E(\tilde{w})$). In particular, the entire diagram commutes up to images.

Next, we note that if we follow the arrows from $\tilde{B}_{i_u}$ along the top of the diagram, and down the right side, then the image of the corresponding composition is exactly $E(\phi_f(w))$, by definition. The image of $\tilde{B}_{i_u}$ as we trace the left side and bottom of the diagram is $(f \circ f_{B_u})(\tilde{B}_{i_u}) = f(B_u) = f(E(w))$. This proves the Proposition in the case that $w \in \mathcal{L}_{red}$.

Now we assume only that $w \in \mathcal{L}$ and $\hat{w}$ is a prefix of w. We let w_{red} denote the (unique) reduced element in the equivalence class of w modulo $\dashrightarrow$. We note that, as we rewrite w, all of the reductions are made to a suffix that does not include any part of the prefix $\hat{w}$, since $\hat{w}$ contains no symbols of the form $[f]$. It follows, in particular, that $w = \hat{w}w'$ and $w_{red} = \hat{w}w''$, and that w'' is the reduced form of w'. Applying ϕ_f to w and w_{red}, we get

$$\phi_f(w) = \tilde{w}[f]w', \qquad \phi_f(w_{red}) = \tilde{w}[f]w''.$$

It follows that

$$\phi_f(w) = \tilde{w}[f]w' \to \tilde{w}[f]w'' = \phi_f(w_{red}).$$

Using the fact that E is constant on equivalence classes modulo $\to$, and the fact that $E(\phi_f(w)) = f(E(w))$ for reduced words w, we see that

$$E(\phi_f(w)) = E(\phi_f(w_{red})) = f(E(w_{red})) = f(E(w)).$$

□

5.5 Witness Automata

Definition 5.38 Let $\widehat{B} \subseteq X$ be a metric ball; choose a finite partition $\mathcal{P}$ of X into open balls such that $\widehat{B} \in \mathcal{P}$. We now define a PDA, called a $\widehat{B}$-*witness automaton*. There are four states: L (the initial state, or *loading* state), R (the *ready* state), C (the *cleaning* state), and E (the *eject* state). The directed edges are as follows.

1 Two types of directed edges lead away from L. The first type is a loop at L having the label $(\epsilon, \epsilon, A_{i,n})$ (i and n range over all possibilities: $i \in \{1, \ldots, k\}$ and $n \in \{1, \ldots, \ell\}$, where k and ℓ are as in Convention 5.19.) There is just one edge of the second type: it leads to the ready state R. Its label is $(\epsilon, \epsilon, addr(\widehat{B}))$.

2 Let $s \in S$, and let B be a region for s. By definition, $s_{|B} = f$, for some $f \in \mathrm{Sim}_X(B, f(B))$. We create a directed edge from R to C with the label

$$(s, addr(B), addr(f(B))[f]);$$

there is one such edge for each $s \in S$ and region B for s.

3 The cleaning state C is the initial vertex for three kinds of edge. First, we note that there is obviously a uniform bound K on the lengths of the words in $\{addr(B)\}$, where the $addr(B)$ are the middle coordinates of the labels of edges leading from the ready state R. For each unreduced word $w \in \Sigma^*$ that occurs as a prefix of length less than or equal to $K+1$ to a word in $\mathcal{L}$, we add a directed loop at C with label $(\epsilon, w, r(w))$, where $r(w)$ is the reduced form of w.

Next, for each ball $\widetilde{B} \in \mathcal{P} - \{\widehat{B}\}$, we add an edge $(\epsilon, addr(\widetilde{B}), \epsilon)$ leading from C to E.

Finally, there is also an edge labeled $(\epsilon, \epsilon, \epsilon)$, leading from C back to R.

4 The edges leading away from state E are all of the same type. They are loops with label (ϵ, A, ϵ), where A is an arbitrary symbol from the alphabet Σ, including the initial stack symbol, $\#$.

Remark With a bit more care, it is possible to specify edges leading away from the loading state L in such a way that it is impossible to arrive in the state R with anything other than a valid word of $\mathcal{L}_{red}$ written on the stack; we will assume that this extra care has been taken, leaving details to the reader.

Proposition 5.39 *For any ball $\widehat{B} \subseteq X$,*

$$\mathcal{L}_{\widehat{B}} = \{w \in S^* \mid w(\widehat{B}) \not\subseteq \widehat{B}\}$$

is the language accepted by a $\widehat{B}$-witness automaton. In particular, $\mathcal{L}_{\widehat{B}}$ is (non-deterministic) context-free.

Proof Let $w \in \mathcal{L}_{\widehat{B}}$. We will prove that w is accepted by a $\widehat{B}$-witness automaton.

We regard $w = s_1 \ldots s_n$ as an element of G. By continuity of w, there is some ball $B \subseteq \widehat{B}$ such that $w(B) \subseteq \widetilde{B}$, for some $\widetilde{B} \in \mathcal{P} - \{\widehat{B}\}$. We may furthermore assume (as in Sublemma 5.27(1)) that $s_j \ldots s_n(B)$ lies inside a region for s_{j-1}, for $j = 2, \ldots, n+1$.

Our automaton begins in state L, with $\#$ written on its stack. It begins by writing the address of B on its stack, and (in the process) moving to state R. Since B lies inside a region D_{s_n} for s_n by our assumptions, it follows that the address for D_{s_n} is a prefix of the address for B. Let $f \in \mathrm{Sim}_X(D_{s_n}, f(D_{s_n}))$ satisfy $s_{n|D_{s_n}} = f$. It follows that we are permitted to follow the directed edge labeled $(s_n, addr(D_{s_n}), addr(f(D_{s_n}))[f])$ (and in fact can follow no other) to state C. We note that, after doing so, the stack value of the path is $\phi_f(addr(B))$. It follows, in particular, that

$$E(\phi_f(addr(B))) = f(E(addr(B))) = s_n(E(addr(B))) = s_n(B),$$

where the first equality is due to Proposition 5.37, the second is due to the equality $f_{|B} = s_{n|B}$, and the third is by the definitions of E and $addr$. It follows that the stack value is a word in the language $\mathcal{L}$ whose reduced form in $\mathcal{L}_{red}$ is the address of $s_n(B)$.

Next, beginning at state C, we repeatedly apply all possible reductions to prefixes of length $K+1$, using the directed loops at C. The effect of doing this is to gather all letters of the form $[f]$ at the end of the prefix (in the $(K+1)$st position at worst; the symbols $[f]$ drop out entirely if the empty stack symbol becomes visible to the automaton). After doing this, we follow the directed edge labeled $(\epsilon, \epsilon, \epsilon)$ back to the ready state R. Note that the stack is "clean" – there are no symbols of the form $[f]$ among the first K symbols on the stack, and, in view of the fact that $s_n(B)$ lies inside a region for s_{n-1}, we can (as above) follow a unique directed edge back to the state C.

The process repeats. Eventually the automaton winds up in state C with a word $w \in \mathcal{L}$ on the stack satisfying

$$E(w) = s_1 \dots s_n(B),$$

and nothing left on the input tape. We again apply the cleaning procedure as described above, resulting in a word w' which still evaluates to $s_1 \dots s_n(B)$, but now has a prefix of length K that is free of the symbols $[f]$. (If the word w' has total length less than K, then w' is entirely free of the symbols $[f]$.) In view of the fact that $s_1 \dots s_n(B) \subseteq \widetilde{B}$ by our assumption, it now follows that the address of $\widetilde{B}$ is a prefix of w'. We may therefore follow the arrow labeled $(\epsilon, addr(\widetilde{B}), \epsilon)$ to the eject state E, where the automaton can completely unload its stack using the directed loops at E. Since the entire input tape has been read and the stack is now empty, the automaton accepts w.

Now let us suppose that $w = s_1 \dots s_n \notin \mathcal{L}_{\widehat{B}}$. We must show that the automaton cannot accept w. The automaton is forced to begin by loading the address of an (unknown) subball B of $\widehat{B}$ on its stack. After doing this, it is in the ready state R. Assuming that the automaton has at least (and, therefore, exactly) one edge to follow from R, it arrives in state C with a word w' written on its stack, such that $E(w')$ is $s_n(B)$. We can then assume that the automaton follows the cleaning procedure sketched above. (Not doing so would only make the automaton less likely to accept w.) At this point, the automaton can move back to the ready state, or (if applicable) to the eject state. However, assuming that $n > 1$, moving to the eject state will cause the automaton to fail, since, from E, there is no longer any opportunity to read the input tape. If $n = 1$ (i.e., if $w = s_n$), then $s_n(B) \subseteq \widehat{B}$, so that, for any $\widetilde{B} \in \mathcal{P} - \{\widehat{B}\}$ the address of $\widetilde{B}$ is not a prefix of the address for $s_n(B)$, and therefore no directed edge from C to E can be crossed.

We may therefore assume that the automaton moves back and forth between the ready and cleaning states, ultimately ending in the cleaning state C with a word w' on the stack, satisfying

$$E(w') = s_1 \dots s_n(B),$$

and no letters on the input tape. We may assume, moreover, that w' has no symbol of the form $[f]$ among its first K entries. Now, since $s_1 \ldots s_n(B) \subseteq \widehat{B}$, it is not possible to follow the directed edge into E. The automaton's only move is to follow the arrow labeled $(\epsilon, \epsilon, \epsilon)$ back to R, where it gets stuck. And hence, it follows that the automaton cannot accept w. □

5.6 Proof of the Main Theorem

Proof of Main Theorem By Proposition 5.26, there is a finite test partition $\mathcal{P}$ for G. We let $\mathcal{P} = \{B_1, \ldots, B_\alpha\}$, where each of the B_i is a metric ball.

Consider the language

$$\hat{\mathcal{L}} = \{w \in S^* \mid w(B_i) \nsubseteq B_i,\, i \in \{1, \ldots, \alpha\}\} = \bigcup_{i=1}^{\alpha} \mathcal{L}_{B_i}.$$

By Proposition 5.39, and because a finite union of context-free languages is context-free, $\hat{\mathcal{L}}$ is context-free.

For any language $\mathcal{L}$, we let $\mathcal{L}^\circ$ denote the cyclic shift of $\mathcal{L}$. That is,

$$\mathcal{L}^\circ = \{w_2 w_1 \in S^* \mid w_1 w_2 \in \mathcal{L};\, w_1, w_2 \in S^*\}.$$

A theorem of [13] says that the cyclic shift of a context-free language is context-free. It follows that $\hat{\mathcal{L}}^\circ$ is context-free.

Finally, we claim that $\mathrm{CoWP}_S(G) = \hat{\mathcal{L}}^\circ$. The reverse direction follows from the (obvious) fact that $\hat{\mathcal{L}} \subseteq \mathrm{CoWP}_S(G)$, and from the fact that the co-word problem is closed under the cyclic shift. Now suppose that $w = s_1 \ldots s_n \in \mathrm{CoWP}_S(G)$. Since $\mathcal{P}$ is a test partition, we must have a ball $B_i \in \mathcal{P}$ such that $s_\beta \ldots s_n s_1 \ldots s_{\beta-1}(B_i) \nsubseteq B_i$. This implies that $w = s_1 \ldots s_n \in \hat{\mathcal{L}}^\circ$.

□

Acknowledgments

The author would like to thank the referee for numerous comments that have helped improve the exposition.

References

[1] Anīsīmov, A. V. 1971. The group languages. *Kibernetika (Kiev)*, 18–24.

[2] Bennett, Daniel and Bleak, Collin. 2016. A dynamical definition of f.g. virtually free groups. *Internat. J. Algebra Comput.*, **26**(1), 105–121.

[3] Berns-Zieve, Rose, Fry, Dana, Gillings, Johnny, Hoganson, Hannah and Mathews, Heather. *Groups with context-free co-word problem and embeddings into Thompson's group V*. This volume. Preprint available at `http://arxiv.org/abs/1407.7745`.

[4] Bleak, Collin and Salazar-Díaz, Olga. 2013. Free products in R. Thompson's group V. *Trans. Amer. Math. Soc.*, **365**(11), 5967–5997.

[5] Bleak, Collin, Matucci, Francesco and Neunhöffer, Max. 2016. Embeddings into Thompson's group V and $co\mathcal{CF}$ groups. *J. Lond. Math. Soc. (2)*, **94**(2), 583–597.

[6] Brookshear, J. Glenn. 1989. *Formal Languages, Automata, and Complexity.* Theory of Computation. The Benjamin/Cummings Publishing Company.

[7] Cannon, J. W., Floyd, W. J. and Parry, W. R. 1996. Introductory notes on Richard Thompson's groups. *Enseign. Math. (2)*, **42**(3-4), 215–256.

[8] Farley, Daniel S. and Hughes, Bruce. 2015. Finiteness properties of some groups of local similarities. *Proc. Edinb. Math. Soc. (2)*, **58**(2), 379–402.

[9] Holt, Derek F., Rees, Sarah, Röver, Claas E. and Thomas, Richard M. 2005. Groups with context-free co-word problem. *J. London Math. Soc. (2)*, **71**(3), 643–657.

[10] Hughes, Bruce. 2009. Local similarities and the Haagerup property. *Groups Geom. Dyn.*, **3**(2), 299–315. With an appendix by Daniel S. Farley.

[11] Lehnert, J. 2008. *Gruppen von quasi-Automorphismen.* Goethe Universität, Frankfurt.

[12] Lehnert, J. and Schweitzer, P. 2007. The co-word problem for the Higman-Thompson group is context-free. *Bull. Lond. Math. Soc.*, **39**(2), 235–241.

[13] Maslov, A. N. 1973. The cyclic shift of languages. *Problemy Peredači Informacii*, **9**(4), 81–87.

[14] Muller, David E. and Schupp, Paul E. 1983. Groups, the theory of ends, and context-free languages. *J. Comput. System Sci.*, **26**(3), 295–310.

[15] Nekrashevych, Volodymyr V. 2004. Cuntz-Pimsner algebras of group actions. *J. Operator Theory*, **52**(2), 223–249.

[16] Newman, M. H. A. 1942. On theories with a combinatorial definition of "equivalence". *Ann. of Math. (2)*, **43**, 223–243.

[17] Röver, Claas E. 1999. Constructing finitely presented simple groups that contain Grigorchuk groups. *J. Algebra*, **220**(1), 284–313.

[18] Shapiro, Michael. 1994. A note on context-sensitive languages and word problems. *Internat. J. Algebra Comput.*, **4**(4), 493–497.

[19] Witzel, Stefan and Zaremsky, Matt. *Thompson groups for systems of groups, and their finiteness properties.* To appear in *Groups Geom. Dyn.* Preprint available at `http://arxiv.org/abs/1405.5491`.

6

Compacta with Shapes of Finite Complexes: a Direct Approach to the Edwards–Geoghegan–Wall Obstruction

Craig R. Guilbault

Abstract

An important "stability" theorem in shape theory, due to D. A. Edwards and R. Geoghegan, characterizes those compacta having the same shape as a finite CW complex. In this chapter we present a straightforward and self-contained proof of that theorem.

6.1 Introduction

Before Ross Geoghegan turned his attention to the main topic of this book, *Topological Methods in Group Theory*, he was a leader in the area of shape theory. In fact, much of his pioneering work in geometric group theory has involved taking key ideas from shape theory and recasting them in the service of groups. Some of his early thoughts on that point of view are captured nicely in [9]. Among the interesting ideas found in that 1986 paper is an early recognition that a group boundary is well-defined up to shape – an idea later formalized by Bestvina in [1].

In this chapter we return to the subject of Geoghegan's early work. For those whose interests lie primarily in group theory, the work presented here contains a concise and fairly gentle introduction to the ideas of shape theory, via a careful study of one of its foundational questions.

In the 1970s D. A. Edwards and R. Geoghegan solved two open problems in shape theory – both related to the issue of "stability". Roughly speaking, these problems ask when a "bad" space has the same shape as a "good" space. For simplicity, we focus on the following versions of those problems.

Problem A. *Give necessary and sufficient conditions for a connected finite-dimensional compactum Z to have the pointed shape of a CW complex.*

Problem B. *Give necessary and sufficient conditions for a connected finite-dimensional compactum Z to have the pointed shape of a finite CW complex.*

Solutions to these problems can be found in the sequence of papers [5], [4], [6]. A pair of particularly nice versions of those solutions are as follows.

Solution A. *Z has the pointed shape of a CW complex if and only if each of its homotopy pro-groups is stable.*

Solution B. *Z has the pointed shape of a finite CW complex if and only if each of its homotopy pro-groups is stable and an intrinsically defined Wall obstruction $\omega(Z,z) \in \widetilde{K}_0(\mathbb{Z}[\check{\pi}_1(Z,z)])$ vanishes.*

Solution B was obtained by combining Solution A with C. T. C. Wall's famous work on finite homotopy types [13]. So, in order to understand Edwards and Geoghegan's solution to Problem B, it is necessary to understand two things: Solution A; and Wall's work on the finiteness obstruction. Since both tasks are substantial – and since Problem B can arise quite naturally without regards to Problem A – we became interested in finding a simpler and more direct solution to Problem B. This chapter contains such a solution. This chapter may be viewed as a sequel to [8], where Geoghegan presented a new and more elementary solution to Problem A. In the same spirit, we feel that our work offers a simplified view of Problem B.

The strategy we use in attacking Problem B is straightforward and very natural. Given a connected n-dimensional pointed compactum Z, begin with an inverse system $K_0 \xleftarrow{f_1} K_1 \xleftarrow{f_2} K_2 \xleftarrow{f_3} \cdots$ of finite n-dimensional (pointed) complexes with (pointed) cellular bonding maps that represents Z. Under the assumption that pro-π_k is stable for all k, we borrow a technique from [7] allowing us to attach cells to the K_is so that the bonding maps induce π_k-isomorphisms for increasingly large k. Our goal then is to reach a *finite* stage where the bonding maps induce π_k-isomorphisms for all k, and are therefore homotopy equivalences. This would imply that Z has the shape of any of those homotopy equivalent finite complexes. As expected, we confront an obstruction lying in the reduced projective class group of pro-π_1. Instead of invoking theorems from [13], we uncover this obstruction in the natural context of the problem at hand; in fact, the main result of [13] can then be obtained as a corollary. Another advantage to the approach taken here is that all CW complexes used in this chapter are finite. This makes both the algebra and the shape theory more elementary.

6.2 Background

In this section we provide some background information on inverse systems, inverse sequences, and shape theory. In addition, we will review the definition of a reduced projective class group. A more complete treatment of inverse systems and sequences can be found in [10]; an expanded version of this introduction can be found in [11]

6.2.1 Inverse Systems

We start with a brief discussion of general inverse systems and pro-categories, which provide the broad framework for more concrete constructions that will follow. A thorough treatment of this topic can be found in [10, Chapter 11].

An *inverse system* $\{X_\alpha, f_\alpha^\beta; \mathcal{A}\}_{\alpha\in\mathcal{A}}$ consists of a collection of objects X_α from a category $\mathcal{C}$ indexed by a *directed set* $\mathcal{A}$, along with morphisms $f_\alpha^\beta : X_\beta \to X_\alpha$ for every pair $\alpha, \beta \in \mathcal{A}$ with $\alpha \leq \beta$, satisfying the property that $f_\alpha^\gamma = f_\beta^\gamma \circ f_\alpha^\beta$ whenever $\alpha \leq \beta \leq \gamma$. By fixing $\mathcal{C}$, but allowing the directed set to vary, and formulating an appropriate definition of morphisms, one obtains a category pro-$\mathcal{C}$ whose objects are all such inverse systems. When $\mathcal{A}' \subseteq \mathcal{A}$ is a directed set there is an obvious subsystem $\{X_\alpha, f_\alpha^\beta; \mathcal{A}'\}_{\alpha\in\mathcal{A}'}$ and an inclusion morphism. When $\mathcal{A}'$ is *cofinal* in $\mathcal{A}$ (for every $\alpha \in \mathcal{A}$ there exists $\alpha' \in \mathcal{A}'$ such that $\alpha \leq \alpha'$), the inclusion morphism is an isomorphism in pro-$\mathcal{C}$. A key theme in this subject is that, when $\mathcal{A}'$ is cofinal, the corresponding subsystem contains all relevant information.

When $\mathcal{C}$ is a category of sets and functions, we may define the *inverse limit* of $\{X_\alpha, f_\alpha^\beta; \mathcal{A}\}_{\alpha\in\mathcal{A}}$ by

$$\varprojlim \{X_\alpha, f_\alpha^\beta; \mathcal{A}\} = \left\{ (x_\alpha) \in \prod_{\alpha\in\mathcal{A}} X_\alpha \,\middle|\, f_\alpha^\beta(x_\beta) = x_\alpha \text{ for all } \alpha \leq \beta \right\}$$

along with projections $p_\alpha : \varprojlim \{X_\alpha, f_\alpha^\beta; \mathcal{A}\} \to X_\alpha$. When $\mathcal{C}$ is made up of topological spaces and maps, the inverse limits are topological spaces and the projections are continuous. Similarly, additional structure is passed along to inverse limits when $\mathcal{C}$ consists of groups, rings, or modules and corresponding homomorphisms. An important example of the "key theme" noted in the previous paragraph is that, when $\mathcal{A}'$ is is cofinal in $\mathcal{A}$, the canonical inclusion $\varprojlim \{X_\alpha, f_\alpha^\beta; \mathcal{A}'\} \to \varprojlim \{X_\alpha, f_\alpha^\beta; \mathcal{A}'\}$ is a bijection of sets [resp., homeomorphism of spaces, isomorphism of groups, etc.].

An *inverse sequence* (or *tower*) is an inverse system for which $\mathcal{A} = \mathbb{N}$, the natural numbers. Since all inverse systems used in this chapter contain cofinal inverse sequences, we are able to work almost entirely with towers. General inverse systems play a useful, but mostly invisible, background role.

6.2.2 Inverse Sequences (aka Towers)

The fundamental notions that make up a category pro-$\mathcal{C}$ are simpler and more intuitive when restricted to the subcategory of towers in $\mathcal{C}$. For our purposes, an understanding of towers will suffice; so that is where we focus our attention.

Let

$$C_0 \xleftarrow{\lambda_1} C_1 \xleftarrow{\lambda_2} C_2 \xleftarrow{\lambda_3} \cdots$$

be an inverse sequence in pro-$\mathcal{C}$. A *subsequence* of $\{C_i, \lambda_i\}$ is an inverse sequence of the form

$$C_{i_0} \xleftarrow{\lambda_{i_0+1} \circ \cdots \circ \lambda_{i_1}} C_{i_1} \xleftarrow{\lambda_{i_1+1} \circ \cdots \circ \lambda_{i_2}} C_{i_2} \xleftarrow{\lambda_{i_2+1} \circ \cdots \circ \lambda_{i_3}} \cdots .$$

In the future we will denote a composition $\lambda_i \circ \cdots \circ \lambda_j$ $(i \leq j)$ by $\lambda_{i,j}$.

Remark *Using the notation introduced in the previous subsection, a bonding map* λ_i *would be labeled by* λ_{i-1}^i *and a composition* $\lambda_i \circ \cdots \circ \lambda_j$ $(i \leq j)$ *by* λ_{i-1}^j. *When working with inverse sequences, we opt for the slightly simpler notation described here.*

Inverse sequences $\{C_i, \lambda_i\}$ and $\{D_i, \mu_i\}$ are isomorphic in pro-$\mathcal{C}$, or *pro-isomorphic*, if after passing to subsequences, there exists a commuting *ladder diagram*

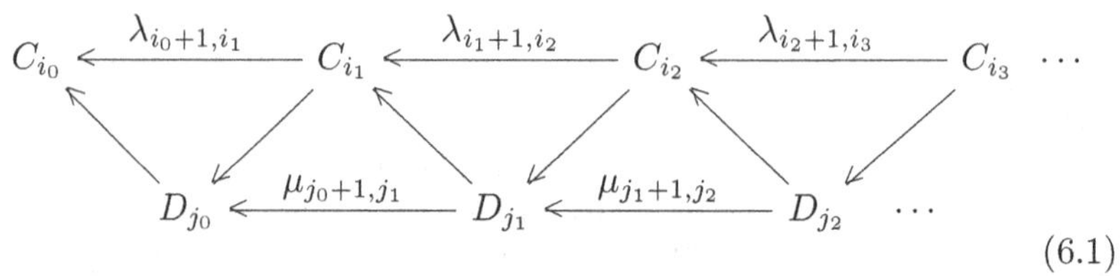

$$(6.1)$$

where the up and down arrows represent morphisms from $\mathcal{C}$. Clearly an inverse sequence is pro-isomorphic to any of its subsequences. To avoid tedious notation, we frequently do not distinguish $\{C_i, \lambda_i\}$ from its subsequences. Instead we simply assume that $\{C_i, \lambda_i\}$ has the desired properties of a preferred subsequence – often prefaced by the words "after passing to a subsequence and relabeling".

Remark *Together the collection of down arrows in (6.1) determine a morphism in* pro-C *from* $\{C_i, \lambda_i\}$ *to* $\{D_i, \mu_i\}$ *and the up arrows a morphism from* $\{D_i, \mu_i\}$ *to* $\{C_i, \lambda_i\}$. *Again see [10, Chapter 11] for details.*

An inverse sequence $\{C_i, \lambda_i\}$ is *stable* if it is pro-isomorphic to a constant sequence

$$D \xleftarrow{\text{id}} D \xleftarrow{\text{id}} D \xleftarrow{\text{id}} \cdots .$$

For example, if each λ_i is an isomorphism from $\mathcal{C}$, it is easy to show that $\{C_i, \lambda_i\}$ is stable.

Inverse limits of inverse sequences of sets are particularly easy to understand. In particular,

$$\varprojlim \{C_i, \lambda_i\} = \left\{ (c_0, c_1, c_2, \cdots) \in \prod_{i=0}^{\infty} C_i \;\middle|\; \lambda_i(c_i) = c_{i-1} \text{ for all } i \geq 1 \right\},$$

with a projection map $p_i : \varprojlim \{C_i, \lambda_i\} \to C_i$ for each $i \geq 0$.

6.2.3 Inverse Sequences of Groups

Of particular interest to us is the category $\mathcal{G}$ of groups and group homomorphisms. It is easy to show that an inverse sequence of groups $\{G_i, \lambda_i\}$ is stable if and only if, after passing to a subsequence and relabeling, there is a commutative diagram of the form

$$\begin{array}{ccccccccc} G_0 & \xleftarrow{\lambda_1} & G_1 & \xleftarrow{\lambda_2} & G_2 & \xleftarrow{\lambda_3} & G_3 & \cdots & \\ \nwarrow & \swarrow & \nwarrow & \swarrow & \nwarrow & \swarrow & & & \\ \operatorname{Im}(\lambda_1) & \xleftarrow{\cong} & \operatorname{Im}(\lambda_2) & \xleftarrow{\cong} & \operatorname{Im}(\lambda_3) & \xleftarrow{\cong} & \cdots & & \end{array}$$

where all unlabeled maps are inclusions or restrictions. In this case $\varprojlim \{C_i, \lambda_i\} \cong \operatorname{im}(\lambda_i)$ and each of the projection homomorphisms takes $\varprojlim \{C_i, \lambda_i\}$ isomorphically onto the corresponding $\operatorname{im}(\lambda_i)$.

The sequence $\{G_i, \lambda_i\}$ is *semistable* (or *pro-epimorphic*, or *Mittag-Leffler*) if it is pro-isomorphic to an inverse sequence $\{H_i, \mu_i\}$ for which each μ_i is surjective. Equivalently, $\{G_i, \lambda_i\}$ is semistable if, after passing to a subsequence and relabeling, there is a commutative diagram of the form

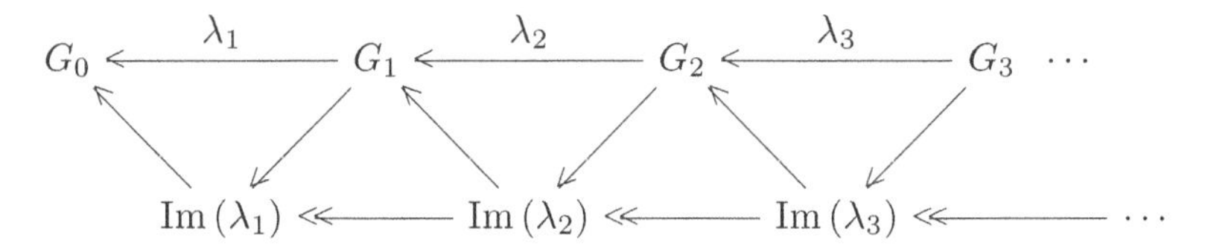

where "$\twoheadleftarrow$" denotes a surjection.

6.2.4 Inverse Systems and Sequences of CW Complexes

Another category of utmost interest to us is $\mathcal{FH}_0$, the category of pointed, connected, finite CW complexes and pointed homotopy classes of maps. (A space is *pointed* if a basepoint has been chosen; a map is *pointed* if basepoint is taken to basepoint.) We will frequently refer to pointed spaces and maps without explicitly mentioning the basepoints. We will refer to an inverse system [resp., tower] from $\mathcal{FH}_0$ as an *inverse system* [resp., *tower*] *of finite complexes.*

For each $k \geq 1$, there is an obvious functor from pro-$\mathcal{FH}_0$ to pro-$\mathcal{G}$ taking an inverse system $\{K_\alpha, g_\alpha^\beta; \Omega\}$ of pointed, connected, finite simplicial complexes to be the inverse system of groups $\{\pi_k(K_\alpha), (g_\alpha^\beta)_*; \Omega\}$ (the kth *homotopy pro-group of* $\{K_\alpha, g_\alpha^\beta; \Omega\}$). A related functor takes $\{K_\alpha, g_\alpha^\beta; \Omega\}$ to the group $\varprojlim \{\pi_k(K_\alpha), (g_\alpha^\beta)_*; \Omega\}$ which we denote by $\check{\pi}_k\left(\{\pi_k(K_\alpha), (g_\alpha^\beta)_*; \Omega\}\right)$ (the kth *Čech homotopy group of* $\{K_\alpha, g_\alpha^\beta; \Omega\}$).

Clearly the initial functor described above takes towers from pro-$\mathcal{FH}_0$ to towers in pro-$\mathcal{G}$, while the latter takes each tower to a group.

6.2.5 Homotopy Dimension

The *dimension*, $\dim(\{K_i, f_i\})$, of a tower of finite complexes is the supremum (possibly ∞) of the dimensions of the K_is. The *homotopy dimen-*

sion of $\{K_i, f_i\}$ is defined by:

$$\begin{aligned}&\operatorname{hom}\dim(\{K_i, f_i\})\\&\quad = \inf\{\dim(\{L_i, g_i\}) \mid \{L_i, g_i\} \text{ is pro-isomorphic to } \{K_i, f_i\}\}.\end{aligned}$$

6.2.6 Shapes of Compacta

Our view of *shape theory* is that it is the study of (possibly bad) compact metric spaces through the use of associated inverse systems and sequences of finite complexes.

Let Z be a compact, connected, metric space with basepoint z. Let Ω denote the set of all finite open covers $\mathcal{U}_\alpha$ of Z, each with a distinguished element U^* containing z. Declare $\mathcal{U}_\alpha \leq \mathcal{U}_\beta$ to mean that $\mathcal{U}_\beta$ refines $\mathcal{U}_\alpha$. Using Lebesgue numbers, it is easy to see that Ω is a directed set. For each $\mathcal{U}_\alpha$, let N_α be its nerve, and for each $\mathcal{U}_\alpha \leq \mathcal{U}_\beta$ let $g_\alpha^\beta : N_\beta \to N_\alpha$ be (the pointed homotopy class of) an induced simplicial map. In this way, we associate with Z an inverse system $\{N_\alpha, g_\alpha^\beta; \Omega\}$ from pro-$\mathcal{FH}_0$. We may then define pro-$\pi_k(Z)$ (the kth *pro-homotopy group of* Z) to be the inverse system $\{\pi_1(N_\alpha), (g_\alpha^\beta)_*; \Omega\}$ and $\check{\pi}_k(Z)$ (the kth *Čech homotopy group of* Z) its inverse limit.

Any cofinal tower contained in the above inverse system will be called a *tower of finite complexes associated with* Z. Another application of Lebesgue numbers shows that such towers always exist. We say that Z and Z' have the same *pointed shape* if their associated towers are pro-isomorphic. The *shape dimension* of Z is defined to be the homotopy dimension of an associated tower. It is easy to see that the shape dimension of Z is less than or equal to its topological dimension.[1]

Since associated towers $\{N_i, g_i\}$ for Z are, by definition, cofinal subsystems of $\{N_\alpha, g_\alpha^\beta; \Omega\}$, each comes with a canonical isomorphism

$$\begin{aligned}&j : \check{\pi}_k(\{N_i, g_i\})\\&\quad = \varprojlim\{\pi_k(N_i), (g_i)_*\} \to \varprojlim\{\pi_k(N_\alpha), (g_\alpha^\beta)_*; \Omega\} \equiv \check{\pi}_k(Z).\end{aligned}$$

6.2.7 The Reduced Projective Class Group

If Λ is a ring, we say that two finitely generated projective Λ-modules P and Q are *stably equivalent* if there exist finitely generated free Λ-

[1] Another method for associating a tower of finite complexes with Z is to realize Z as the inverse limit of such complexes. It is a standard fact in shape theory that such a sequence will be pro-isomorphic to the ones obtained above. See, for example, [2] or [12].

modules F_1 and F_2 such that $P \oplus F_1 \cong Q \oplus F_2$. Under the operation of direct sum, the stable equivalence classes of finitely generated projective modules form a group $\widetilde{K}_0(\Lambda)$, known as the *reduced projective class group* of Λ. In this group, a finitely generated projective Λ-module P represents the trivial element if and only if it is *stably free*, i.e., there exists a finitely generated free Λ-module F such that $P \oplus F$ is free.

Of particular interest is the case where G is a group and Λ is the group ring $\mathbb{Z}[G]$. Then $\widetilde{K}_0$ determines a functor from the category $\mathcal{G}$ of groups to the category $\mathcal{AG}$ of abelian groups. In particular, a group homomorphism $\lambda : G \to H$ induces a ring homomorphism $\mathbb{Z}[G] \to \mathbb{Z}[H]$, which induces a group homomorphism $\lambda_* : \widetilde{K}_0(\mathbb{Z}[G]) \to \widetilde{K}_0(\mathbb{Z}[H])$.

6.3 Main Results

We are now ready to state and prove the main results of this chapter.

Theorem 6.1 *Let $\{K_i, f_i\}$ be a finite-dimensional tower of pointed, connected, finite complexes having stable* pro-π_k *for all k. Then there is a well defined obstruction $\omega(\{K_i, f_i\}) \in \widetilde{K}_0(\mathbb{Z}[\check{\pi}_1(\{K_i, f_i\})])$ which vanishes if and only if $\{K_i, f_i\}$ is stable in* pro-$\mathcal{FH}_0$.

Translating Theorem 6.1 into the language of shape theory yields the desired solution to Problem B.

Theorem 6.2 *A connected compactum Z with finite shape dimension has the pointed shape of a finite CW complex if and only if each of its homotopy pro-groups is stable and an intrinsically defined Wall obstruction $\omega(Z) \in \widetilde{K}_0(\mathbb{Z}[\check{\pi}_1(Z)])$ vanishes.*

Our proof of Theorem 6.1 begins with two lemmas. The first is a simple and well-known algebraic observation.

Lemma 6.3 *Let C_* be a chain complex of finitely generated free Λ-modules, and suppose that $H_i(C_*) = 0$ for $i \leq k$. Then*

1 $\ker \partial_i$ *is finitely generated and stably free for all $i \leq k+1$, and*
2 $H_{k+1}(C_*)$ is finitely generated.

Proof For the first assertion, begin by noting that $\ker \partial_0 = C_0$ is finitely generated and free. Proceeding inductively for $j \leq k+1$, assume that

$\ker \partial_{j-1}$ is finitely generated and stably free. Since $H_{j-1}(C_*)$ is trivial, we have a short exact sequence

$$0 \to \ker \partial_j \to C_j \to \ker \partial_{j-1} \to 0.$$

By our assumption on $\ker \partial_{j-1}$, the sequence splits. Therefore, $\ker \partial_j \oplus \ker \partial_{j-1} \cong C_j$, which implies that $\ker \partial_j$ is finitely generated and stably free.

The second assertion follows from the first since by definition we have $H_{k+1}(C_*) = \ker \partial_{k+1} / \operatorname{im} \partial_{k+2}$. □

The second lemma – which is really the starting point to our proof of Theorem 6.1 – was extracted from [7, Theorem 4]. It uses the following standard notation and terminology. For a map $f : K \to L$, the mapping cylinder of f will be denoted $M(f)$. The relative homotopy and homology groups of the pair $(M(f), K)$ will be abbreviated to $\pi_i(f)$ and $H_i(f)$. We say that f is *k-connected* if $\pi_i(f) = 0$ for all $i \leq k$; or equivalently, $f_* : \pi_i(K) \to \pi_i(L)$ is an isomorphism for $i < k$ and a surjection when $i = k$. The universal cover of a space K will be denoted $\widetilde{K}$. If $f : K \to L$ induces a π_1-isomorphism, then $\widetilde{f} : \widetilde{K} \to \widetilde{L}$ denotes a lift of f.

Lemma 6.4 (The Tower Improvement Lemma) *Let $\{K_i, f_i\}$ be a tower of pointed, connected, finite complexes with stable* pro-π_k *for $k \leq n$ and semistable* pro-π_{n+1}. *Then there is a pro-isomorphic tower $\{L_i, g_i\}$ of finite complexes with the property that each g_i is $(n+1)$-connected. Moreover, after passing to a subsequence of $\{K_i, f_i\}$ and relabeling, we may assume that:*

1. *each L_i is constructed from K_i by inductively attaching finitely many k-cells for $2 \leq k \leq n+2$,*
2. *each g_i is an extension of f_i with $g_i(K_i \cup (\text{new cells of dimension} \leq k)) \subset (K_{i-1} \cup (\text{new cells of dimension } \leq k-1))$, and*
3. *the inclusions $K_i \hookrightarrow L_i$ form the promised pro-isomorphism from $\{K_i, f_i\}$ to $\{L_i, g_i\}$.*

Proof Our proof is by induction on n.

Step 1. $(n = 0)$ Let $\{K_i, f_i\}$ be a tower with semistable pro-π_1. By attaching 2-cells to the K_is, we wish to obtain a new tower in which all bonding maps induce surjections on π_1.

By semistability, we may (by passing to a subsequence and relabeling) assume that each f_{i*} maps $f_{i+1*}(\pi_1(K_{i+1}))$ onto $f_{i*}(\pi_1(K_i))$. Let

$\{{}^i a_j\}_{j=1}^{N_i}$ be a finite generating set for $\pi_1(K_i)$ and for each ${}^i a_j$ choose ${}^i b_j \in f_{i+1*}(\pi_1(K_{i+1}))$ such that $f_{i*}({}^i a_j) = f_{i*}({}^i b_j)$. For each element of the form ${}^i a_j ({}^i b_j)^{-1} \in \pi_1(K_i)$, attach a 2-cell to K_i which kills that element. Call the resulting complexes L_is, and note that each f_i extends to a map $k_i : L_i \to K_{i-1}$. Define $g_i : L_i \to L_{i-1}$ to be k_i composed with the inclusion $K_{i-1} \hookrightarrow L_{i-1}$. This leads to the following commutative diagram:

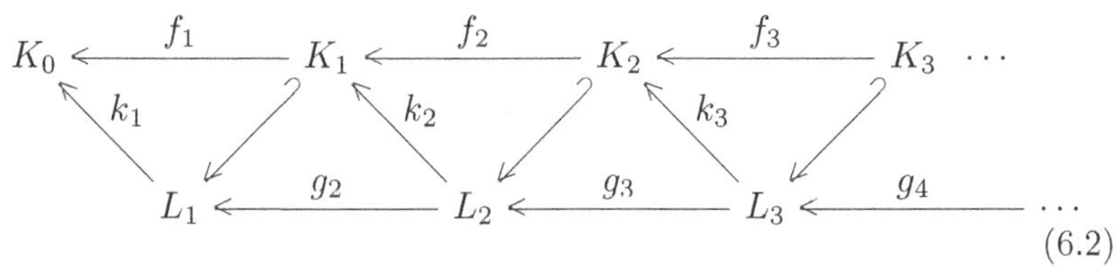

(6.2)

which ensures that the tower $\{L_i, g_i\}$ is pro-isomorphic to the original via inclusions.

Note that each $g_{i+1*} : \pi_1(L_{i+1}) \to \pi_1(L_i)$ is surjective. Indeed, the loops in K_i corresponding to the generating set $\{{}^i a_j\}$ of $\pi_1(K_i)$ still generate $\pi_1(L_i)$; moreover, in $\pi_1(L_i)$ each ${}^i a_j$ becomes identified with ${}^i b_j$ which lies in $\operatorname{im}(g_{i+1*})$. Properties 1 and 2 are immediate from the construction.

Step 2. ($n > 0$) Now suppose $\{K_i, f_i\}$ is a tower such that pro-π_k is stable for all $k \leq n$ and pro-π_{n+1} is semistable.

We may assume inductively that there is a tower $\{L'_i, g'_i\}$ which has n-connected bonding maps and (after passing to a subsequence of $\{K_i, f_i\}$ and relabeling) satisfies:

1′ each L'_i is constructed from K_i by inductively attaching finitely many k-cells for $2 \leq k \leq n+1$,

2′ each g'_i is an extension of f_i such that

$$g'_i\,(K_i \cup (\text{new cells of dimension } \leq k))$$
$$\subset (K_{i-1} \cup (\text{new cells of dimension } \leq k-1))\,,$$

3′ $\{L'_i, g'_i\}$ and $\{K_i, f_i\}$ are pro-isomorphic via inclusions.

Since pro-π_{n+1} is semistable, we may also assume that:

4′ g'_{i*} maps $g'_{i+1*}(\pi_{n+1}(L'_{i+1}))$ onto $g'_{i*}(\pi_{n+1}(L'_i))$ for all i.

Since the g'_is are n-connected, then each $g'_{i*} : \pi_k(L'_i) \to \pi_k(L'_{i-1})$ is an isomorphism for $k < n$. In addition, each $g'_{i*} : \pi_n(L'_i) \to \pi_n(L'_{i-1})$

is surjective; but since pro-π_n is stable, all but finitely many of these surjections must be isomorphisms. So, by dropping finitely many terms and relabeling, we assume that these also are isomorphisms.

Our goal is now clear – by attaching $(n+2)$-cells to the L_i's, we wish to make each bonding map $(n+1)$-connected.

Due to the π_n-isomorphisms just established, we have an exact sequence

$$\cdots \to \pi_{n+1}\left(L_i'\right) \xrightarrow{g_{i*}'} \pi_{n+1}\left(L_{i-1}'\right) \to \pi_{n+1}\left(g_i'\right) \to 0, \tag{6.3}$$

for each i. Furthermore, since $n \geq 1$, each g_i' induces a π_1-isomorphism, so we may pass to the universal covers to obtain (by covering space theory and the Hurewicz theorem) isomorphisms:

$$\pi_{n+1}\left(g_i'\right) \cong \pi_{n+1}\left(\widetilde{g}_i'\right) \cong H_{n+1}\left(\widetilde{g}_i'\right). \tag{6.4}$$

Each term in the cellular chain complex $C_*\left(\widetilde{g}_i'\right)$ is a finitely generated $\mathbb{Z}[\pi_1\left(L_i\right)]$-module; so, by Lemma 6.3, $H_{n+1}\left(\widetilde{g}_i'\right)$ is finitely generated.

Applying (6.4), we may choose a finite generating set $\left\{{}^i\bar{\alpha}_j\right\}_{j=1}^{N_i}$ for each $\pi_{n+1}\left(g_i'\right)$; and by (6.3), each ${}^i\bar{\alpha}_j$ may be represented by an ${}^i\alpha_j' \in \pi_{n+1}\left(L_{i-1}'\right)$. By Condition 3′ we may choose for each ${}^i\alpha_j'$, some ${}^i\beta_j \in \pi_{n+1}\left(L_i'\right)$ such that $g_{i-1}' \circ g_i'\left({}^i\beta_j\right) = g_{i-1}'\left({}^i\alpha_j'\right)$. Let ${}^i\alpha_j = {}^i\alpha_j' - g_i'\left({}^i\beta_j\right) \in \pi_{n+1}\left(L_{i-1}'\right)$. Then each ${}^i\alpha_j$ is sent to ${}^i\bar{\alpha}_j$ in $\pi_{n+1}\left(g_i'\right)$ and $g_{i-1*}'\left({}^i\alpha_j\right) = 0 \in \pi_{n+1}\left(L_{i-2}'\right)$. Attach $(n+2)$-cells to each L_{i-1}' to kill the ${}^i\alpha_j$s. Call the resulting complexes L_is, and for each i let $k_i : L_i \to L_{i-1}'$ be an extension of g_i'. Then let $g_i : L_i \to L_{i-1}$ be the composition of k_i with the inclusion $L_{i-1}' \hookrightarrow L_{i-1}$. This leads to a diagram like that produced in Step 1, hence the new system $\{L_i, g_i\}$ is pro-isomorphic to $\{L_i', g_i'\}$, and thus to $\{K_i, f_i\}$ via inclusions. Moreover, it is easy to check that each g_i is $(n+1)$-connected. Properties 1 and 2 are immediate from the construction and the inductive hypothesis, and Property 3 from the final ladder diagram. □

Suppose now that $\{K_i, f_i\}$ has stable pro-π_k for all k. Then, by repeatedly attaching cells to the K_is, one may obtain pro-isomorphic towers with r-connected bonding maps for arbitrarily large r. If $\{K_i, f_i\}$ is finite-dimensional it seems reasonable that, once r exceeds the dimension of $\{K_i, f_i\}$, this procedure will terminate with bonding maps that are connected in all dimensions – and thus, homotopy equivalences. Unfortunately, this strategy is too simplistic – in order to obtain r-connected maps we must attach $(r+1)$-cells; thus, the dimensions of the complexes continually exceed the connectivity of the bonding maps.

Roughly speaking, Theorem 6.1 captures the obstruction to making this strategy work.

Proof of Theorem 6.1 Begin with a tower $\{L_i, g_i\}$ of q-dimensional complexes pro-isomorphic to $\{K_i, f_i\}$, via a diagram of type (6.2), which has the following properties for all i.

(a) g_i is $(q-1)$-connected,
(b) for $k \in \{q-2, q-1, q\}$, g_i maps the k-skeleton of L_i into the $(k-1)$-skeleton of L_{i-1},
(c) g_{i*} maps $g_{i+1*}(\pi_q(L_{i+1}))$ onto $g_{i*}(\pi_q(L_i))$, and
(d) $g_{i*} : \pi_{q-1}(L_i) \to \pi_{q-1}(L_{i-1})$ is an isomorphism.

A tower satisfying Conditions (a) and (b) is easily obtainable; apply Lemma 6.4 to $\{K_i, f_i\}$ with $n = \dim\{K_i, f_i\} + 1$, in which case $q = \dim\{K_i, f_i\} + 3$. (**Note.** Although it may seem excessive to allow the $\dim\{L_i, g_i\}$ to exceed $\dim\{K_i, f_i\}$ by 3, this is done to obtain Condition (b), which is key to our argument.) Semistability of pro-π_q gives Condition (c) – after passing to a subsequence and relabeling. Then, since pro-π_{q-1} is stable and each $g_{i*} : \pi_{q-1}(L_i) \to \pi_{q-1}(L_{i-1})$ is surjective, we may drop finitely many terms to obtain Condition (d).

As in the proof of Lemma 6.4, $\pi_q(g_i)$ and $H_q(\widetilde{g}_i)$ are isomorphic finitely generated $\mathbb{Z}[\pi_1 L_i]$-modules. We will show that, for all i, $H_q(\widetilde{g}_i)$ is projective and that all of these modules are stably equivalent. (This is a pleasant surprise, since the L_is and g_is may all be different.) Thus we obtain corresponding elements $[H_q(\widetilde{g}_i)]$ of $\widetilde{K}_0(\mathbb{Z}[\pi_1(L_i)])$. When these elements are trivial, i.e., when the modules are stably free, we will show that, by attaching finitely many $(q+1)$-cells to each L_i, bonding maps can be made homotopy equivalences. To complete the proof we define a single obstruction $\omega(\{K_i, f_i\})$ to be the image of $(-1)^{q+1}[H_q(\widetilde{g}_i)]$ in $\widetilde{K}_0(\mathbb{Z}[\check{\pi}_1(\{K_i, f_i\})])$ and show that this element is uniquely determined by $\{K_i, f_i\}$.

Notes (1) To be more precise, the elements $H_q(\widetilde{g}_i)$ determine the elements $(-1)^{q+1}[H_q(\widetilde{g}_i)]$ of the group $\widetilde{K}_0(\mathbb{Z}[\pi_1(L_i)])$ which may be associated, via inclusion maps (that induce π_1-isomorphisms), with elements of $\widetilde{K}_0(\mathbb{Z}[\pi_1(K_i)])$, which in turn determine a common element of the limit group $\widetilde{K}_0(\mathbb{Z}[\check{\pi}_1(\{K_i, f_i\})])$ via the projection maps – which in our setting are all isomorphisms.

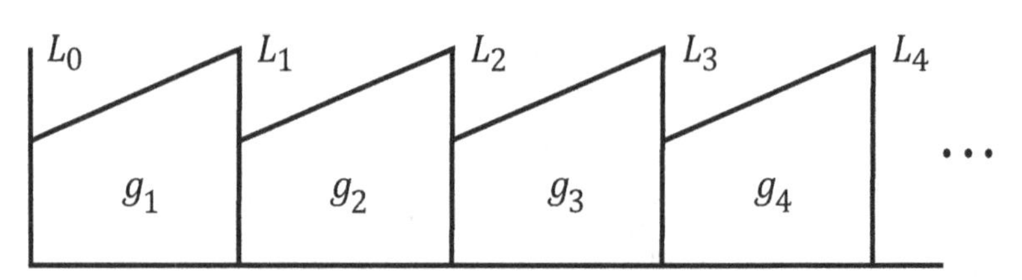

Figure 6.1

(2) We have used a factor $(-1)^{q+1}$ (instead of the more concise $(-1)^q$) so that our definition agrees with those already in the literature.

While most of our work takes place in the individual mapping cylinders $M(g_i)$ and their universal covers, there is some interplay between adjacent cylinders. For that reason, it is useful to view our work as taking place in the "infinite mapping telescope" shown in Figure 6.1 (and in its universal cover).

For ease of notation, fix i and consider the pair $(M(\widetilde{g}_i), \widetilde{L}_i)$. It is a standard fact (see [3, 3.9]) that $C_*(\widetilde{g}_i)$ is isomorphic to the algebraic mapping cone of the chain homomorphism $g_{i*} : C_*(\widetilde{L}_i) \to C_*(\widetilde{L}_{i-1})$. In particular, if the cellular chain complexes $C_*(\widetilde{L}_{i-1})$ and $C_*(\widetilde{L}_i)$ of $\widetilde{L}_{i-1}$ and $\widetilde{L}_i$ are expressed as:

$$0 \to D_q \xrightarrow{d_q} D_{q-1} \xrightarrow{d_{q-1}} \cdots \xrightarrow{d_2} D_1 \xrightarrow{d_1} D_0 \longrightarrow 0 \quad \text{and} \tag{6.5}$$

$$0 \to D'_q \xrightarrow{d'_q} D'_{q-1} \xrightarrow{d'_{q-1}} \cdots \xrightarrow{d'_2} D'_1 \xrightarrow{d'_1} D'_0 \longrightarrow 0, \tag{6.6}$$

respectively, then $C_*(\widetilde{g}_i)$ is naturally isomorphic to a chain complex

$$0 \to C_{q+1} \xrightarrow{\partial_{q+1}} C_q \xrightarrow{\partial_q} \cdots \xrightarrow{\partial_2} C_1 \xrightarrow{\partial_1} C_0 \longrightarrow 0,$$

where, for each j,

$$C_j = D'_{j-1} \oplus D_j \quad \text{and} \quad \partial_j(x, y) = (-d'_{j-1}x,\ \widetilde{g}_{i*}x + d_j y).$$

Here one views each $\pi_1(L_{i-1})$-module D_j as a $\pi_1(L_i)$-module in the obvious way – associating $a \cdot x$ with $\widetilde{g}_{i*}(a) \cdot x$ for $a \in \pi_1(L_i)$.

By Condition (b), the map $\widetilde{g}_{i*} : D'_j \to D_j$ is trivial for $j \geq q-2$; so, in these dimensions, ∂_j splits as $-d'_{j-1} \oplus d_j$, allowing our chain complex to be written:

$$0 \to \underbrace{D'_q \oplus 0}_{C_{q+1}} \overset{-d'_q \oplus 0}{\longrightarrow} \underbrace{D'_{q-1} \oplus D_q}_{C_q} \overset{-d'_{q-1} \oplus d_q}{\longrightarrow} \underbrace{D'_{q-2} \oplus D_{q-1}}_{C_{q-1}}$$

$$\overset{-d'_{q-2} \oplus d_{q-1}}{\longrightarrow} \underbrace{D'_{q-3} \oplus D_{q-2}}_{C_{q-2}} \overset{\partial_{q-2}}{\longrightarrow} \cdots .$$

Since the "minus signs" have no effect on kernels or images of maps, it follows that

$$\ker \partial_{q-1} = \ker(d'_{q-2}) \oplus \ker(d_{q-1}), \tag{6.7}$$

$$\ker \partial_q = \ker(d'_{q-1}) \oplus \ker(d_q), \tag{6.8}$$

$$\ker \partial_{q+1} = \ker(d'_q), \tag{6.9}$$

$$H_{q-1}(\widetilde{g}_i) = \left(\ker(d'_{q-2}) / \operatorname{im}\left(d'_{q-1}\right)\right) \oplus \left(\ker(d_{q-1}) / \operatorname{im}\left(d_q\right)\right), \tag{6.10}$$

$$H_q(\widetilde{g}_i) = \left(\ker(d'_{q-1}) / \operatorname{im}\left(d'_q\right)\right) \oplus \ker(d_q), \tag{6.11}$$

$$H_{q+1}(\widetilde{g}_i) = \ker(d'_q). \tag{6.12}$$

Since $H_{q-1}(\widetilde{g}_i) = 0$, each summand in Identity (6.10) is trivial. Furthermore, the same reasoning applied to the adjacent mapping cylinder $M(g_{i+1})$ yields an analogous set of identities for $C_*(\widetilde{g}_{i+1})$ in which the "primed terms" become the "unprimed terms". This shows that $\ker(d'_{q-1}) / \operatorname{im}\left(d'_q\right)$ is also trivial. Hence, the first summand in Identity (6.11) is trivial, so $H_q(\widetilde{g}_i) \cong \ker(d_q)$. Identity (6.8) together with Lemma 6.3 then shows that $H_q(\widetilde{g}_i)$ is finitely generated and projective. The same reasoning in $C_*(\widetilde{g}_{i+1})$ shows that $H_q(\widetilde{g}_{i+1}) \cong \ker(d'_q)$ is finitely generated projective, so by Identity (6.12), $H_{q+1}(\widetilde{g}_i)$ is finitely generated projective and naturally isomorphic to $H_q(\widetilde{g}_{i+1})$ (using $\widetilde{g}_{i+1*}$ to make $H_{q+1}(\widetilde{g}_i)$ a $\pi_1(L_{i+1})$-module).

Next we show that $H_q(\widetilde{g}_i)$ and $H_{q+1}(\widetilde{g}_i)$ are stably equivalent. Extract the short exact sequence

$$0 \to H_{q+1}(\widetilde{g}_i) \to D'_q \to \operatorname{im}\left(d'_q\right) \to 0$$

from above, then recall that $\operatorname{im}\left(d'_q\right)$ is equal to $\ker(d'_{q-1})$. The latter is projective, so

$$D'_q \cong H_{q+1}(\widetilde{g}_i) \oplus \ker(d'_{q-1}).$$

Thus $[H_{q+1}(\widetilde{g}_i)] = -\left[\ker(d'_{q-1})\right]$ in $\widetilde{K}_0(\pi_1(L_i))$. Combining the fact that $[H_q(\widetilde{g}_i)] = [\ker(d_q)]$ and, by Identity (6.8), $[\ker(d_q)] = -\left[\ker(d'_{q-1})\right]$, we obtain $[H_q(\widetilde{g}_i)] = [H_{q+1}(\widetilde{g}_i)]$.

To summarize, we have shown that for each i:

- $H_q(\widetilde{g}_i)$ and $H_{q+1}(\widetilde{g}_i)$ are finitely generated and projective,
- $[H_q(\widetilde{g}_i)] = [H_{q+1}(\widetilde{g}_i)]$ in $\widetilde{K}_0(\pi_1(L_i))$, and
- $H_{q+1}(\widetilde{g}_i)$ is naturally isomorphic to $H_q(\widetilde{g}_{i+1})$ as $\pi_1(L_{i+1})$-modules.

These observations combine to show that each $[H_q(\widetilde{g}_i)]$ determines the "same" element of $\widetilde{K}_0(\check{\pi}_1(\{K_i, f_i\}))$. More precisely, define $\omega(\{K_i, f_i\})$ to be the image of $(-1)^{q+1}[H_q(\widetilde{g}_i)]$ under the isomorphism

$$\widetilde{K}_0(\mathbb{Z}[\pi_1(L_i)]) \to \widetilde{K}_0(\check{\pi}_1(\{K_i, f_i\}))$$

induced by the composition of group isomorphisms

$$\pi_1(L_i) \xrightarrow{p_i^{-1}} \varprojlim \{\pi_i(L_i), (g_i)_*\} \to \varprojlim \{\pi_i(K_i), (f_i)_*\} = \check{\pi}_1(\{K_i, f_i\}), \tag{6.13}$$

where $p_i : \varprojlim \{\pi_i(L_i), (g_i)_*\} \to \pi_1(L_i)$ is the projection map, and the isomorphism between inverse limits is canonically induced by ladder diagram (6.2).

Claim 1. If $\omega(\{K_i, f_i\}) = 0$, then $\{K_i, f_i\}$ is stable.

We will show that, by adding finitely many q- and $(q+1)$-cells to each of the above L_is, we may arrive at a pro-isomorphic tower in which all bonding maps are homotopy equivalences.

By assumption, each $\mathbb{Z}\pi_1$-module $H_q(\widetilde{g}_i)$ becomes free upon summation with a finitely generated free module. This may be accomplished geometrically by attaching finitely many q-cells to the corresponding L_{i-1}s via trivial attaching maps at the basepoints. Each g_{i-1} may then be extended by mapping these q-cells to the basepoint of L_{i-2}. Since this procedure preserves all relevant properties of our tower, we will assume that, for each i, $H_q(\widetilde{g}_i)$ (and therefore $\pi_q(g_i)$) is a finitely generated free $\mathbb{Z}[\pi_1(L_i)]$-module.

Proceed as in Step 2 of the proof of Lemma 6.4 to obtain collections $\{{}^i\alpha_j\}_{j=1}^{N_i} \subset \pi_{n+1}(L_{i-1})$ that correspond to generating sets for the $\pi_q(g_i)$s and which satisfy $g_{i-1*}({}^i\alpha_j) = 0 \in \pi_q(L_{i-2})$ for all i, j. In addition, we now require that $\{{}^i\alpha_j\}_{j=1}^{N_i}$ corresponds to a free basis for $\pi_q(g_i)$. For each ${}^i\alpha_j$ attach a single $(q+1)$-cell to L_{i-1} to kill that element. Extend each g_i to $g_i' : L_i' \to L_{i-1}'$ as before, thereby obtaining a tower $\{L_i', g_i'\}$ for which all bonding maps are q-connected. Since the $(q+1)$-cells are attached to L_{i-1} along a free basis, we do not create any new $(q+1)$-cycles for the pair $\left(M(\widetilde{g}_i'), \widetilde{L}_i'\right)$, so no new

$(q+1)$-dimensional homology is introduced. Moreover, the $(q+1)$-cells attached to L_{i-1} result in $(q+2)$-cells in $M(\widetilde{g}'_{i-1})$ which are attached in precisely the correct manner to kill $H_{q+1}(\widetilde{g}_{i-1})$ without creating any $(q+2)$-dimensional homology – this is due to the natural isomorphism discovered earlier between $H_{q+1}(\widetilde{g}_{i-1})$ and $H_q(\widetilde{g}_i)$. Thus the g'_is are all $(n+2)$-connected, and since the L'_is are $(n+1)$-dimensional, this means that the g'_is are homotopy equivalences. So $\{L'_i, g'_i\}$ and hence $\{K_i, f_i\}$, are stable in pro-$\mathcal{FH}_0$.

Claim 2. The obstruction is well defined.

We must show that $\omega(\{K_i, f_i\})$ does not depend on the tower $\{L_i, g_i\}$ and ladder diagram chosen at the beginning of the proof. First observe that any subsequence $\{L_{k_i}, g_{k_i k_{i-1}}\}$ of $\{L_i, g_i\}$ yields the same obstruction. This is immediate in the special case that $\{L_{k_i}, g_{k_i k_{i-1}}\}$ contains two consecutive terms of $\{L_i, g_i\}$. If not, notice that $\{L_{k_i}, g_{k_i k_{i-1}}\}$ is a subsequence of $L_{k_1} \leftarrow L_{k_1+1} \leftarrow L_{k_2} \leftarrow L_{k_3} \leftarrow \cdots$, which is a subsequence of $\{L_i, g_i\}$. Therefore the more general observation follows from the special case.

Next suppose that $\{L_i, g_i\}$ and $\{M_i, h_i\}$ are each towers of finite q-dimensional complexes satisfying the conditions laid out at the beginning of the proof. Then $\{L_i, g_i\}$ and $\{M_i, h_i\}$ are pro-isomorphic; so, after passing to subsequences and relabeling, there exists a homotopy commuting diagram of the form:

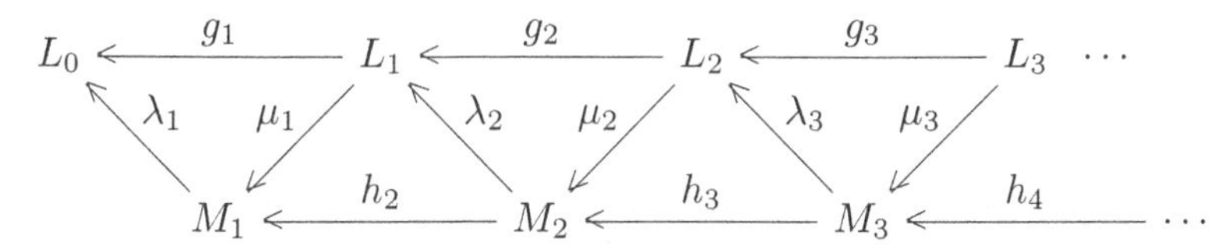

where all λ_i and μ_i are cellular maps. From here we may create a new tower:

$$M_1 \longleftarrow L_2 \longleftarrow M_4 \longleftarrow L_5 \longleftarrow M_7 \longleftarrow L_8 \longleftarrow M_{10} \longleftarrow \cdots$$

where the bonding maps are determined (up to homotopy) by the ladder diagram. Properties (a), (c) and (d) hold for this tower due to the corresponding properties for $\{L_i, g_i\}$ and $\{M_i, h_i\}$. To see that Property (b) holds, note that each bonding map is the composition of a g_i or an h_i with a cellular map. (This is why so many terms were omitted.) Since this new tower contains subsequences which are – up to homotopies of the bonding maps – subsequences of $\{L_i, g_i\}$ and $\{M_i, h_i\}$, our initial observation implies that all determine the same obstruction.

Finally we consider the general situation where $\{L_i, g_i\}$ and $\{M_i, h_i\}$ satisfy Conditions (a)-(d), but are not necessarily of the same dimension. By the previous case and induction, it will be enough to show that, for a given q-dimensional $\{L_i, g_i\}$, we can find a $(q+1)$-dimensional tower $\{L'_i, g'_i\}$ which satisfies the corresponding versions of Conditions (a)-(d), and which determines the same obstruction as $\{L_i, g_i\}$. In this step, the need for the $(-1)^{q+1}$ factor finally becomes clear.

The tower $\{L'_i, g'_i\}$ is obtained by carrying out our usual strategy of attaching a finite collection of $(q+1)$-cells to each L_{i-1} along a generating set for $H_q(M(\widetilde{g}_i), \widetilde{L}_i)$. The resulting $C_*(\widetilde{L}'_i)$s differ from the $C_*(\widetilde{L}_i)$s only in dimension $q+1$ where we have introduced finitely generated free modules ${}^iF_{q+1}$. By inserting this term into (6.5) and rewriting D_q as $\operatorname{im}(d_q) \oplus \ker(d_q)$, the chain complex for L'_{i-1} may be written:

$$0 \longrightarrow {}^iF_{q+1} \xrightarrow{d_{q+1}} \operatorname{im}(d_q) \oplus \ker(d_q) \xrightarrow{d_q} D_{q-1} \xrightarrow{d_{q-1}} \\ \cdots \xrightarrow{d_2} D_1 \xrightarrow{d_1} D_0 \longrightarrow 0.$$

By construction, d_{q+1} takes ${}^iF_{q+1}$ onto $\ker(d_q)$ thereby eliminating the q-dimensional homology of the pair $(M(\widetilde{g}'_i), \widetilde{L}'_i)$. Note, however, that we may have introduced new $(q+1)$-dimensional homology. Indeed, by our earlier analysis, $H_{q+1}(\widetilde{g}'_i) = \ker(d_{q+1})$. (The original $(q+1)$-dimensional homology of the pair was eliminated – as it was in the unobstructed case – when we attached $(q+1)$-cells to L_i.) By extracting the short exact sequence

$$0 \longrightarrow \ker(d_{q+1}) \longrightarrow {}^iF_{q+1} \longrightarrow \ker(d_q) \longrightarrow 0$$

and recalling that $\ker(d_q) \cong H_q(\widetilde{g}_i)$ is projective, we have

$${}^iF_{q+1} \cong H_{q+1}(\widetilde{g}'_i) \oplus H_q(\widetilde{g}_i).$$

So, upon projection into $\widetilde{K}_0(\check{\pi}_1(\{K_i, f_i\}))$, (as described in equation (6.13)), $[H_{q+1}(\widetilde{g}'_i)]$ and $-[H_q(\widetilde{g}_i)]$ determine the same element. The same is then true for $(-1)^{q+2}[H_{q+1}(\widetilde{g}'_i)]$ and $(-1)^{q+1}[H_q(\widetilde{g}_i)]$, showing that $\{L_i, g_i\}$ and $\{L'_i, g'_i\}$ lead to the same obstruction. □

Proof of Theorem 6.2 We need only verify the forward implication, as the converse is obvious.

Using the finite-dimensionality of Z, choose a finite-dimensional tower of pointed, connected, finite complexes $\{N_i, g_i\}$ associated with Z. By the pro-π_k hypotheses on Z, we may apply Theorem 6.1 to obtain $\omega(\{N_i, g_i\}) \in \widetilde{K}_0(\mathbb{Z}[\check{\pi}_1(\{N_i, g_i\})]))$. The inclusion of $\{N_i, g_i\}$ into the associated inverse system $\{N_\alpha, g_\alpha^\beta; \Omega\}$, as described in Subsection 6.2.6, yields a canonical isomorphism of $\check{\pi}_1(\{K_i, g_i\})$ onto $\check{\pi}_1(\{N_\alpha, g_\alpha^\beta; \Omega\}) = \check{\pi}_1(Z)$ which converts $\omega(\{N_i, g_i\})$ to our intrinsically defined Wall obstruction $\omega(Z) \in \widetilde{K}_0(\mathbb{Z}[\check{\pi}_1(Z)])$. □

6.4 Realizing the Obstructions

In addition to proving Theorems 6.1 and 6.2, Edwards and Geoghegan showed how to build towers and compacta with non-trivial obstructions. By applying their strategy within our framework, we obtain an easy proof of the following.

Proposition 6.5 *Let G be a finitely presentable group and P a finitely generated projective $\mathbb{Z}[G]$ module. Then there exists a tower of finite 2-complexes $\{K_i, f_i\}$, with stable* pro-π_k *for all k and $\check{\pi}_1(\{K_i, f_i\}) \cong G$, such that $\omega(\{K_i, f_i\}) = [P] \in \widetilde{K}_0(\mathbb{Z}[G])$.*

By letting $Z = \varprojlim \{K_i, f_i\}$ we immediately obtain the following.

Proposition 6.6 *Let G be a finitely presentable group and P a finitely generated projective $\mathbb{Z}[G]$ module. Then there exists a compact connected 2-dimensional pointed compactum Z, with stable* pro-π_k *for all k and $\check{\pi}_1(Z) \cong G$, such that $\omega(Z) = [P] \in \widetilde{K}_0(\mathbb{Z}[G])$.*

Proof Let Q be a finitely generated projective $\mathbb{Z}[G]$ module representing $-[P]$ in $\widetilde{K}_0(\mathbb{Z}[G])$, and so that $F = P \oplus Q$ is finitely generated and free. Let r denote the rank of F. Let K' be a finite pointed 2-complex with $\pi_1(K') \cong G$, then construct K from K' by wedging a bouquet of r 2-spheres to K' at the basepoint. Then $\pi_2(K) \cong H_2\left(\widetilde{K}\right)$ has a summand isomorphic to F which corresponds to the bouquet of 2-spheres. Define a map $f : K \to K$ so that $f|_{K'} = \text{id}$ and $f_* : \pi_2(K) \to \pi_2(K)$ (or equivalently $\widetilde{f}_* : H_2(\widetilde{K}) \to H_2(\widetilde{K})$) is the projection $P \oplus Q \to P$ when restricted to the F-factor. Note that $H_2(\widetilde{f}) \cong Q \cong H_3(\widetilde{f})$. Obtain the tower $\{K_i, f_i\}$ by letting $K_i = K$ for all $k \geq 0$ and $f_i = f$ for all $k \geq 1$.

To calculate $\omega(\{K_i, f_i\})$ according to the proof of Theorem 6.1, we must attach cells of dimensions 3, 4, and 5 to each K_i to obtain an equivalent tower $\{L_i, g_i\}$ satisfying Conditions (a)-(d) of the proof. As we saw in Claim 2 of Theorem 6.1, this procedure simply shifts homology to higher dimensions. In particular, $[H_5(\widetilde{g}_i)] = -[H_2(\widetilde{f})] = [P]$, as desired. □

Acknowledgment

Work on this project was aided by a Simons Foundation Collaboration Grant.

References

[1] Bestvina, Mladen. 1996. Local homology properties of boundaries of groups. *Michigan Math. J.*, **43**(1), 123–139.

[2] Borsuk, K. 1971. *Theory of shape.* Matematisk Institut, Aarhus Universitet. Lecture Notes Series, No. 28.

[3] Cohen, Marshall M. 1973. *A course in simple-homotopy theory.* Springer-Verlag. Graduate Texts in Mathematics, Vol. 10.

[4] Edwards, David A. and Geoghegan, Ross. 1975a. Shapes of complexes, ends of manifolds, homotopy limits and the Wall obstruction. *Ann. of Math. (2)*, **101**, 521–535.

[5] Edwards, David A. and Geoghegan, Ross. 1975b. The stability problem in shape, and a Whitehead theorem in pro-homotopy. *Trans. Amer. Math. Soc.*, **214**, 261–277.

[6] Edwards, David A. and Geoghegan, Ross. 1976. Stability theorems in shape and pro-homotopy. *Trans. Amer. Math. Soc.*, **222**, 389–403.

[7] Ferry, Steve. 1980. A stable converse to the Vietoris-Smale theorem with applications to shape theory. *Trans. Amer. Math. Soc.*, **261**(2), 369–386.

[8] Geoghegan, Ross. 1978. Elementary proofs of stability theorems in pro-homotopy and shape. *General Topology and Appl.*, **8**(3), 265–281.

[9] Geoghegan, Ross. 1986. The shape of a group – connections between shape theory and the homology of groups. Pages 271–280 of: *Geometric and algebraic topology.* Banach Center Publ., vol. 18. PWN.

[10] Geoghegan, Ross. 2008. *Topological methods in group theory.* Graduate Texts in Mathematics, vol. 243. Springer.

[11] Guilbault, Craig R. 2016. Ends, shapes, and boundaries in manifold topology and geometric group theory. Pages 45–125 of: *Topology and geometric group theory.* Springer Proc. Math. Stat., vol. 184. Springer.

[12] Mardešić, Sibe and Segal, Jack. 1982. *Shape theory.* North-Holland Mathematical Library, vol. 26. North-Holland Publishing Co.

[13] Wall, C. T. C. 1965. Finiteness conditions for CW-complexes. *Ann. of Math. (2)*, **81**, 56–69.

7
The Horofunction Boundary of the Lamplighter Group L_2 with the Diestel–Leader metric

Keith Jones and Gregory A. Kelsey

Abstract

We fully describe the horofunction boundary $\partial_h L_2$ with the word metric associated with the generating set $\{t, at\}$ (i.e. the metric arising in the Diestel–Leader graph DL$(2,2)$). The visual boundary $\partial_\infty L_2$ with this metric is a subset of $\partial_h L_2$. Although $\partial_\infty L_2$ does not embed continuously in $\partial_h L_2$, it naturally splits into two subspaces, each of which is a punctured Cantor set and does embed continuously. The height function on DL$(2,2)$ provides a natural stratification of $\partial_h L_2$, in which countably-many non-Busemann points interpolate between the two halves of $\partial_\infty L_2$. Furthermore, the height function and its negation are themselves non-Busemann horofunctions in $\partial_h L_2$ and are global fixed points of the action of L_2.

7.1 Introduction

The horofunction boundary $\partial_h X$ of a proper complete metric space (X, d) is in general defined as a subspace of the quotient of $C(X)$, the space of continuous $\mathbb{R}$-valued functions on X, by constant functions [1, Definition II.8.12]. It suffices to choose a base point b in X and use the embedding $i : X \hookrightarrow C(X)$ sending $z \in X \mapsto d(z, x) - d(z, b)$. Since X is proper, the closure $\overline{X}$ of $i(X)$ in $C(X)$ provides a compactification of X. We define $\partial_h X$ to be $\overline{X} \backslash i(X)$. We call a point in $\overline{X}$ a *horofunction*, and given a sequence (y_n) of points in X, one can define a horofunction associated with (y_n) by

$$h_{y_n}(x) = \lim_{n\to\infty} d(y_n, x) - d(y_n, b), \tag{7.1}$$

provided this limit exists.

Gromov defines the horofunction boundary, which he calls the *ideal boundary*, in the context of hyperbolic manifolds [5], but the definition applies to any complete metric space. In [1] Bridson and Haefliger use this construction in the context of CAT(0) spaces as a functorial construction of the visual boundary. The horofunction boundary also naturally arises in the study of group C^*-algebras, where Rieffel, referring to it as the *metric boundary*, demonstrates its usefulness particularly in determining the C^*-algebra he calls the *cosphere algebra* [10, Section 3].

In this chapter, X is a group with a word metric, which is $\mathbb{N}$-valued.[1] In this setting, we define a geodesic ray to be an isometric embedding $\mathbb{N} \to X$. We refer to a point of $\partial_h X$ as a *Busemann point* if it corresponds to a sequence of points lying along a geodesic ray. We will refer to the space of asymptotic classes of geodesic rays in (X, d) as the *visual boundary* $\partial_\infty X$. In CAT(0) spaces, all horofunctions correspond to Busemann points; in fact, we can extend i to $\bar{i} : X \sqcup \partial_\infty X \to \overline{X}$, and this is a homeomorphism [1, Section II.8.13]. In general one cannot expect an injective, surjective, or even continuous map from $\partial_\infty X$ to $\partial_h X$. Rieffel brings up the question of determining for a given space (X, d) which points of $\partial_h X$ are Busemann points [10, after Definition 4.8]. As an interesting example of non-injectivity, Reiffel demonstrates that there are no non-Busemann points in $\partial_h \mathbb{Z}^n$ with the ℓ_1 norm, and there are countably many Busemann points [10]. However, Kitzmiller and Rathbun demonstrate that $\partial_\infty \mathbb{Z}^n$ is uncountable [7].

Others have studied the horofunction boundary of Cayley graphs of non-CAT(0) groups, often with variation in their terminology, though examples are still sparse.[2] Develin extended Rieffel's work to abelian groups (he refers to the horofunction boundary as a *Cayley compactification* of the group) [2]. Friedland and Freitas found explicit formulas for horofunctions for $GL(n, \mathbb{C})/\mathrm{U}_n$ with Finsler p-metrics (they use the term *Busemann compactification*) [3]. Webster and Winchester (using the term *metric boundary* as Rieffel) studied the action of a word hyperbolic group on its horofunction boundary and found it is amenable [14]. They also established necessary and sufficient conditions for an infinite graph to have non-Busemann points in its horofunction boundary [15]. Walsh has considered the horofunction boundaries of Artin groups of dihedral type [12] and the action of a nilpotent group on its horo-

[1] For us, $\mathbb{N}$ contains 0.

[2] Note, for non-CAT(0) groups, ∂_h depends on the generating set, as is demonstrated in [10, Example 5.2].

function boundary [13]. Klein and Nicas have studied the horofunction boundary of the Heisenberg group equipped with different metrics [8], [9]. They determine the isometry group of the Heisenberg group with the Carnot–Carathéodory metric.

The *lamplighter group*, discussed more fully at the start of Section 7.2, is given by the presentation:

$$L_2 = \langle a, t \mid a^2, [a^{t^i}, a^{t^j}] \forall\, i, j \in \mathbb{Z} \rangle.$$

Let $S = \{t, at\}$. The generating set S naturally arises when viewing the lamplighter group as a group generated by a finite state automaton (FSA) [4]. This is a rare case where we are able to understand the Cayley graph of such a group with its FSA generating set. In this case, the Cayley graph is the Diestel–Leader graph DL(2, 2) [16]. In [6], the authors describe the visual boundary for Diestel–Leader graphs, which are certain graphs arising from products of regular trees. When there are more than two trees, the topology is indiscrete, but for two trees, the graph inherits enough structure from its component trees that its visual boundary is an interesting non-Hausdorff space. Since DL(2, 2) (the product of two trees with valence 3) is a Cayley graph for L_2, this provides a boundary for L_2 which is dependent on the generating set. This boundary has a natural partition into two uncountable subsets, which we refer to as the upper and lower visual boundaries and denote by $\partial_\infty L_2^+$ and $\partial_\infty L_2^-$. When equipped with the subspace topology, these subsets *are* Hausdorff.

The goal of this chapter is to fully describe $\partial_h L_2$ where the metric on L_2 is the word metric from S. In Section 7.2 we provide some background on this metric, and in Section 7.3 we discuss the relationship between $\partial_\infty L_2$ and $\partial_h L_2$, proving the following theorem.

Theorem (A - Corollary 7.5 and Observations 7.6 and 7.24) *There is a natural map $\partial_\infty L_2 \to \partial_h L_2$, which is injective but not continuous. When restricted to either $\partial_\infty L_2^+$ or $\partial_\infty L_2^-$, however, this injection is continuous.*

In Section 7.4, we explicitly compute formulas for families of horofunctions, including Busemann functions. It turns out the natural height map $H : L_2 \to \mathbb{Z}$ (see Definition 7.1) is a non-Busemann horofunction. Section 7.5 provides a proof that all of the points in $\partial_h L_2$ are members of the families described in Section 7.4, which is our main result.

Theorem (B - Corollary 7.22) *Every point in $\partial_h L_2$ belongs to one of* *the following families of horofunctions, all of whose formulas we explicitly calculate in Section 7.4.*

- *Busemann: these horofunctions arise from certain sequences of lamp stands where the union of positions of lit lamps is bounded below or above and the position of the lamplighter limits to positive or negative infinity.*
- *Spine: these horofunctions arise from certain sequences of lamp stands where the union of lit lamps is bounded neither below nor above and the position of the lamplighter limits to a finite value.*
- *Ribs: these horofunctions arise from certain sequences of lamp stands where the union of positions of lit lamps is bounded below or above but not both and the position of the lamplighter limits to a finite value.*
- *Height: the natural height function and its negation arise as horofunctions from certain sequences of lamp stands where the union of lit lamps is bounded neither below nor above and the position of the lamplighter limits to positive or negative infinity.*

The spine is parametrized by $\mathbb{Z}$ *and the ribs by a subset of* L_2*, and so the set of non-Busemann horofunctions is countable.*

We describe the topology of $\partial_h L_2$ in Section 7.6 by determining the accumulation points, leading to the visualization in Figure 7.1.

The names of the spine and ribs families come from the topology. The spine family is parametrized by the limiting position of the lamplighter and appears in Figure 7.1 as the central column of points. For each spine function, there exist two subfamilies of ribs – a "positive rib" and a "negative rib" – each a countable discrete subspace with the spine function as its only accumulation point. See the discussion in Subsection 7.4.2 for a thorough description of these subfamilies.

Finally, Section 7.7 deals with some properties of the natural action of L_2 on $\partial_h L_2$, in particular noting that $\pm H$ are global fixed points.

7.2 The Diestel–Leader Metric on L_2

Let d denote the word metric on L_2 with generating set $S = \{t, at\}$. Since this is the metric on L_2 induced by the Cayley graph DL(2, 2), we refer to it as the *Diestel–Leader metric* on L_2. Whenever we refer to $\partial_\infty L_2$ or $\partial_h L_2$, we always mean with d. Stein and Taback have calculated

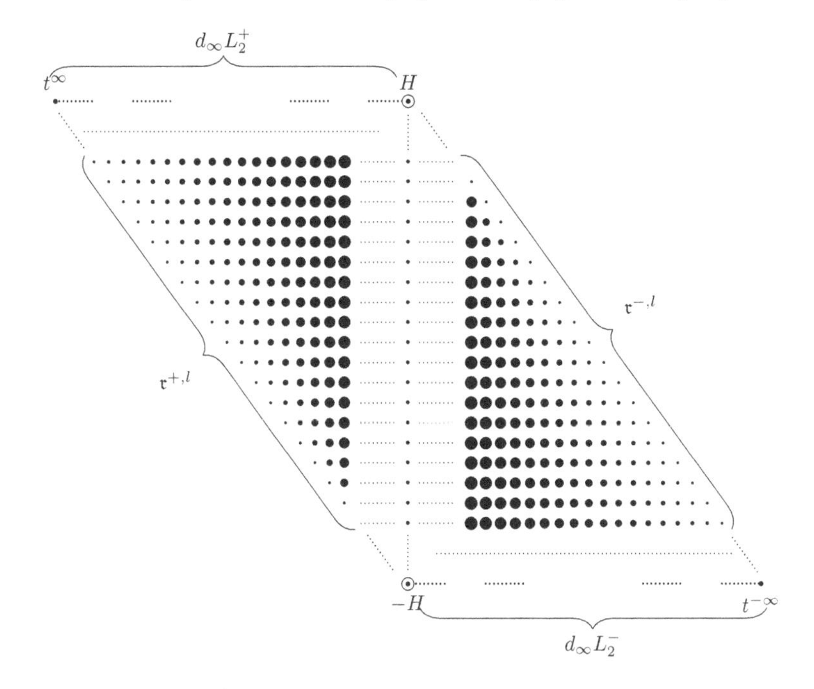

Figure 7.1 Visualization of the horoboundary, including the spine (central column), ribs (discrete point sets limiting to the corresponding spine point), and upper and lower visual boundaries. For each rib, the dots of increasing size represent finite discrete sets whose cardinalities double as we approach the spine.

the metric for general Diestel–Leader graphs [11], but in our case it is simple enough to review and provide a proof.

Each element of L_2 is associated with a "lamp stand", which consists of an infinite row of lamps in bijective correspondence with $\mathbb{Z}$, finitely many of which are lit, and a marked lamp indicating the position of the lamplighter. Figure 7.2 illustrates a typical example. The lamps are binary: either on or off. Right multiplying by a toggles the lamp at the lamplighter's position, while right multiplying by t increments the position of the lamplighter. We think of this increment as a "step right" as in the figure. Using S, the actions are either "step (right or left)" for $t^{\pm 1}$, "toggle then step right" for at, or "step left then toggle" for $(at)^{-1} = t^{-1}a$.

Definition 7.1 For $g \in L_2$, we define $H(g)$ to be the position of

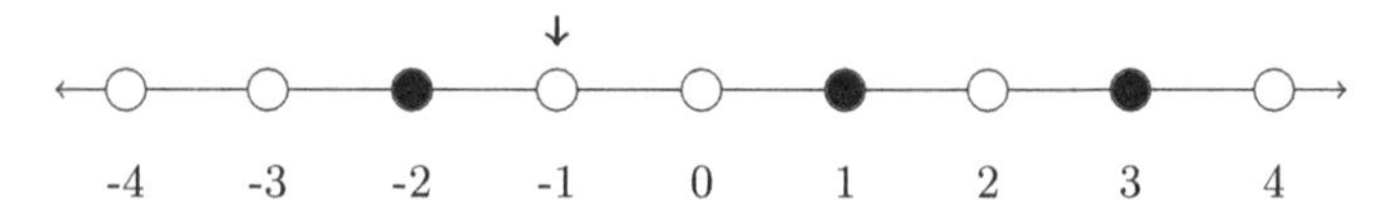

Figure 7.2 A typical element of L_2.

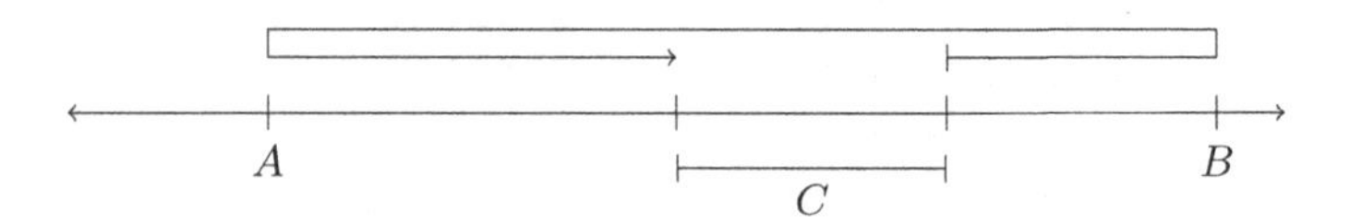

Figure 7.3 Distance between two elements of L_2 with Diestel–Leader metric.

the lamplighter in the lamp stand representing g, or equivalently the exponent sum of t in a word representing g, or the height of g in $\mathrm{DL}(2,2)$.

We define $m(g)$ to be equal to the minimum position of a lit lamp in the lamp stand representation of g if the set of lit lamps is non-empty, and equal to $+\infty$ otherwise. Similarly, we define $M(g)$ to be equal to the maximum position of a lit lamp in the lamp stand representation of g if the set of lit lamps is non-empty, and equal to $-\infty$ otherwise.

For $g_1, g_2 \in L_2$, we define $m(g_1, g_2)$ to be the minimum position of a lamp whose status differs in the lamp stands of g_1 and g_2 if such a position exists, and equal to $+\infty$ otherwise. Similarly, $M(g_1, g_2)$ is the maximum position of a lamp whose status differs in the lamp stands of g_1 and g_2 if such a position exists, and is $-\infty$ otherwise.

We will define "infinite lamp stands" to represent boundary elements. For these lamp stands, we define the H, m, and M notation analogously.

Lemma 7.2 *If $g_1, g_2 \in L_2$, then $d(g_1, g_2) = 2(B - A) - C$ where*

- *$A = \min\{m(g_1, g_2), H(g_1), H(g_2)\}$ is the left-most position the lamplighter must visit to change between g_1 and g_2,*
- *$B = \max\{M(g_1, g_2)+1, H(g_1), H(g_2)\}$ is the right-most such position, and*
- *$C = |H(g_2)-H(g_1)|$ is the distance between the lamplighter's positions in g_1 and g_2.*

See Figure 7.3 for an illustration of a typical path.

Proof Since the Cayley graph is vertex transitive, without loss of generality we may assume that $g_1 = id$ and we denote g_2 simply by g. We will consider a geodesic from id to g on the lamp stand representations of the elements of L_2.

A geodesic will start at id with no lit lamps and the lamplighter at position $H(id) = 0$. The lamplighter must move in one direction (either left or right) until it has gone as far as it needs to, it then travels to the other extremal position, and then finishes by moving to $H(g)$. The initial direction will be *away* from $H(g)$ in order to minimize the total distance. Notice that the minimum extremal position is given by A, which in this case is $A = \min\{m(g), H(g), 0\}$, and the maximal extremal position is given by B, which in this case is $B = \max\{M(g) + 1, H(g), 0\}$. Notice that we use $M(g)+1$ and not $M(g)$ since to turn on the lamp at position k, the lamplighter must be at position $k + 1$ either immediately before turning on lamp k (if using generator $(at)^{-1}$) or immediately after (if using generator at).

Thus, the second of the three segments of the geodesic will have length $B - A$. The lengths of the first and third segments will sum to less than $B - A$, and the amount less will be exactly equal to the distance between the starting and ending position, which in our case is $|H(g)|$. □

7.3 Busemann Points

7.3.1 The Visual Boundary

As in [6, Section 3.3], we can interpret elements of the visual boundary in terms of the lamp stand model. Such an element can be represented by a geodesic ray emanating from the identity which follows a sequence of steps wherein the lamplighter first moves one direction until reaching the extremal lit lamp in that direction then "turns around" and marches off towards $\pm\infty$ toggling lamps as necessary. Thus, in the limit there is either a minimal lit lamp (if any are lit at all), and the lamplighter is at $+\infty$; or there is a maximal lit lamp (if any are lit at all), and the lamplighter is at $-\infty$. A "turning around" only occurs if the minimal lit lamp has negative index in the former case, or the maximal lit lamp has positive index in the latter. This final configuration of lit lamps gives an "infinite lamp stand" for the geodesic ray.

In [6, Observations 4.10 and 4.11] the authors investigate the visual boundary of DL(2, 2) and find that as a set, it is a disjoint union of the sets $\partial_\infty L_2^+$ and $\partial_\infty L_2^-$, where $\partial_\infty L_2^\pm$ is the set of those asymptotic classes with lamplighter at $\pm\infty$. These two sets both have the subset topology of punctured Cantor sets, but the full visual boundary is not Hausdorff. We provide the intuition here.

By [6, Lemma 3.5] in L_2 geodesic rays that are asymptotic eventually merge. For example, if a ray has the lamplighter go from 0 to $-n$ and then in the positive direction forever, the lamps from $-n$ to 0 will be traversed twice. Therefore, the initial setting of lamps on the first pass can be redone on the second pass. The asymptotic class of the ray includes all the different initial settings that become the same final setting when the lamplighter moves in its final direction. Thus, the infinite lamp stand of a ray is actually an invariant of its asymptotic class.

Notice that such a ray that has the lamplighter go from 0 to $-n$ and then in the positive direction forever is in $\partial_\infty L_2^+$, but is close in the visual boundary topology to rays in $\partial_\infty L_2^-$ that have the lamplighter only move in the negative direction and agree on the initial settings of the lamps 0 through $-n$. The fact that these initial settings can be made arbitrary within the asymptotic equivalence class gives us a large subset of $\partial_\infty L_2^-$ that is contained in a neighborhood of *any* ray in $\partial_\infty L_2^+$ where the lamplighter moves in the negative direction for a long time before eventually moving in the positive direction forever.

Thus, there exist distinct elements of $\partial_\infty L_2^+$ whose neighborhoods always intersect, and that intersection is a subset of $\partial_\infty L_2^-$. Therefore $\partial_\infty L_2$ is not Hausdorff. Recall that both $\partial_\infty L_2^\pm$ are punctured Cantor sets under the subspace topology. So, while the subspace topologies of these "halves" are Hausdorff, they are not compact. The full visual boundary $\partial_\infty L_2$ is, however, compact, since these troublesome open sets that intersect both $\partial_\infty L_2^\pm$ "fill" the punctures with open sets in the opposite half.

7.3.2 The Visual Boundary as a Subset of the Horofunction Boundary

We now show that there is a natural injection from the non-Hausdorff $\partial_\infty L_2 = \partial_\infty \mathrm{DL}(2,2)$ into $\partial_h L_2$. Since $\partial_h L_2$ is Hausdorff, this injection is non-continuous.

Lemma 7.3 (Lemma 8.18(1) in Chapter II.8 of [1]) *Let γ be a geodesic ray in $DL(2,2)$ based at the identity. Then the sequence of points $(\gamma(n))$ defines a horofunction $\mathfrak{b}^\gamma$.*

The horofunction $\mathfrak{b}^\gamma$ is called the *Busemann function* associated with

γ. In a CAT(0) space, the Busemann functions of two rays are equal if and only if those two rays are asymptotic. Even though DL(2, 2) is not CAT(0), the same is true in our case.

Lemma 7.4 *Let γ, γ' be geodesic rays in the Cayley graph of L_2 based at the identity. The Busemann functions $\mathfrak{b}^{\gamma}$ and $\mathfrak{b}^{\gamma'}$ are equal if and only if γ and γ' are asymptotic to each other.*

Proof Recall that asymptotic rays in DL(2, 2) eventually merge. Thus, the Busemann functions of asymptotic rays are equal.

Now suppose that γ and γ' are *not* asymptotic to each other. Let $\alpha \in [\gamma], \alpha' \in [\gamma']$ (i.e. α is in the asymptotic equivalence class of γ) so that α and α' have maximal shared initial segment. Say that this shared initial segment has length k. Let $x = \alpha(k+1)$. Notice that by definition, $\mathfrak{b}^{\alpha}(x) = -(k+1)$. By our choice of α and α', $\mathfrak{b}^{\alpha'}(x) = -(k-1)$, so $\mathfrak{b}^{\alpha} \neq \mathfrak{b}^{\alpha'}$. By the proof above of the other direction, $\mathfrak{b}^{\gamma} = \mathfrak{b}^{\alpha}$ and $\mathfrak{b}^{\gamma'} = \mathfrak{b}^{\alpha'}$ and we are done. □

Corollary 7.5 *The relation taking an asymptotic equivalence class of geodesic rays based at the identity to their Busemann functions is an injection of $\partial_\infty L_2$ into $\partial_h L_2$.*

Observation 7.6 *The injection in Corollary 7.5 is not continuous.*

Proof The continuous injective image of a non-Hausdorff space like $\partial_\infty L_2$ must also be non-Hausdorff, while $C(L_2)$ (and thus its subspace $\partial_h L_2$) is Hausdorff. □

Recall that the non-Hausdorff property was proved by finding neighborhoods of distinct elements of $\partial_\infty L_2^+$ that always shared elements of $\partial_\infty L_2^-$. Observation 7.24 shows that the restriction of this injection to either of the subspaces of the visual boundary $\partial_\infty L_2^{\pm}$ is continuous.

7.4 Model Horofunctions

In this section, we construct four families of "model" horofunctions, and in Section 7.5 we show that these represent all horofunctions.

We break $\partial_h L_2$ into four categories: the Busemann points, the *spine*, the *ribs*, and the two points $\pm H$. The reader may refer to Figure 7.1 to preview a visualization of the boundary, illustrating our choice of terms. To determine which category a sequence (x_n) in L_2 falls into (if it defines a horofunction at all), it turns out we need only consider whether $H(x_n)$

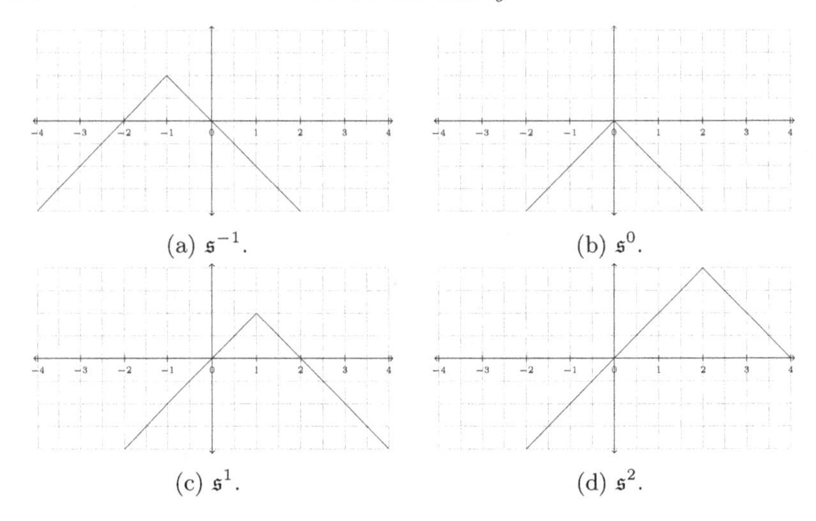

(a) $\mathfrak{s}^{-1}$. (b) $\mathfrak{s}^{0}$. (c) $\mathfrak{s}^{1}$. (d) $\mathfrak{s}^{2}$.

Figure 7.4 Four "spinal" horofunctions, as functions of the height of $g \in L_2$.

approaches an integer or $\pm\infty$, and whether the union over all lit lamps in the sequence is bounded above or below.

7.4.1 The Spine

Fix $l \in \mathbb{Z}$, and let $(s_n^l), n \in \mathbb{N}$, be the sequence in L_2 having lamps at $\pm n$ lit, all others unlit, and $H(s_n^l) = l$.

Applying Lemma 7.2 with $A = -n$, $B = n+1$, and $C = |l|$, we have $d(s_n^l, id) = 4n + 2 - |l|$. Given any $g \in L_2$, take

$$n > \max\{-m(g), M(g), |H(g)|, |l|\},$$

and apply Lemma 7.2 to obtain $d(s_n^l, g) = 4n + 2 - |l - H(g)|$. By Equation 7.1, the horofunction is

$$\mathfrak{s}^l(g) = h_{s_n^l}(g) - |l| - |l - H(g)|. \tag{7.2}$$

We call this the *spine horofunction at height* l. For a given l, this is a function of only $H(g)$. Figure 7.4 shows the graphs of $\mathfrak{s}^{-1}$, $\mathfrak{s}^0$, $\mathfrak{s}^1$, and $\mathfrak{s}^2$, respectively, as functions of height. The spine horofunction at height 0 is $\mathfrak{s}^0 = -|H(g)|$. One can check that the sequence (s_n^n) yields $H(g)$ and (s_n^{-n}) yields $-H(g)$; and we can see that the spine functions interpolate between the two.

7.4.2 The Ribs

The rib horofunctions will be parametrized by certain elements of L_2. There are two subfamilies, corresponding to the $+\infty$ and $-\infty$ direction, and the generating set $\{t, at\}$ creates a slight asymmetry between them. Let $f \in L_2$, and set $l = H(f)$.

First, assume $M(f) < l$ (noting that $M(f)$ may equal $-\infty$). Then consider the sequence $(r_n^{+,f})$, $n \geq l$, in L_2 where the lamps of $r_n^{+,f}$ agree with those of f in each position below l, $H(r_n^{+,f}) = l$, lamp n is lit, and no other lamps at positions l or above are lit.

Given $g \in L_2$, take n large enough, and we have:

$$d(r_n^{+,f}, id) = 2((n+1) - \min\{m(f), l, 0\}) - |l|,$$
$$d(r_n^{+,f}, g) = 2((n+1) - \min\{m(f,g), H(g), l\}) - |l - H(g)|.$$

This yields the *(positive) rib horofunction* corresponding to f:

$$\begin{aligned}\mathfrak{r}^{+,f}(g) &= 2(\min\{m(f), l, 0\} - \min\{m(f,g), H(g), l\}) - |l - H(g)| + |l| \\ &= 2(\min\{m(f), l, 0\} - \min\{m(f,g), H(g), l\}) + \mathfrak{s}^l(g). \qquad (7.3)\end{aligned}$$

We can see that if we had chosen an element whose lamps agreed with f below l, but also had lamps in position l or higher lit, defining a sequence similarly would lead us to the same horofunction, since we can always toggle lamps at l or above "for free" with the generator at.

Though the set of positive rib horofunctions is discrete, there is some structure to be observed. Given a height l, let R_l^+ be the set of positive rib horofunctions at height l. Each corresponds to an element $f \in L_2$ with $H(f) = l$ and $M(f) < l$. Then the "minimum lit lamp" map $m : L_2 \to \overline{\mathbb{Z}}$ induces a map $\hat{m}_l : R_l^+ \to \overline{\mathbb{Z}}$. For $k \in \overline{\mathbb{Z}}$, the cardinality of $\hat{m}_l^{-1}(k)$ is $2^{(l-k-1)}$ if $k < l$, 1 if $k = +\infty$, and 0 otherwise. The set R_l^+ can then be partitioned according to the non-empty pre-images, which provides a natural filtration of R_l^+. Any sequence (r_n) of horofunctions in R_l^+ corresponding to a sequence (f_n) in L_2 with $m(f_n) \to -\infty$, will approach $\mathfrak{s}^l$. We make a precise argument for this fact in Observation 7.26.

In the special case that f has no lit lamps, then $M(f) = -\infty$ and $m(f) = +\infty$ and the calculation simplifies. Since $m(f,g) = m(g)$, the only data is the height $l = H(f)$; and we have:

$$\begin{aligned}\mathfrak{r}^{+,f}(g) = \mathfrak{r}^{+,l}(g) &= 2(\min\{l, 0\} - \min\{m(g), H(g), l\}) - |l - H(g)| + |l| \\ &= -2\min\{m(g), H(g), l\} - |l - H(g)| + l. \qquad (7.4)\end{aligned}$$

As indicated in the preceding paragraph, when there are no lit lamps the resulting horofunction $\mathfrak{r}^{+,l}$ is in a sense the farthest positive rib function of height l from the spine, and we think of it as the *rib tip* at height l.

We now turn to the negative rib functions, corresponding to those f that satisfy $m(f) \geq l$ (possibly with $m(f) = +\infty$). Note we use "$\geq$" now, whereas we used "$<$" previously, since the status of the lamp at l does matter in this direction, since using $(at)^{-1}$ will only let us toggle lamps in positions $l - 1$ or lower "for free". One can define a corresponding sequence similarly to the positive direction, except that the lit lamps approach $-\infty$, and calculate the horofunction $\mathfrak{r}^{-,f}(g)$ to be

$$\begin{aligned} &2(\max\{M(f,g)+1, H(g), l\} - \max\{M(f)+1, l, 0\}) - |l - H(g)| + |l| \\ &\quad = 2(\max\{M(f,g)+1, H(g), l\} - \max\{M(f)+1, l, 0\}) + \mathfrak{s}^{l}(g). \quad (7.5) \end{aligned}$$

There is a similar simplification in this direction when f has no lit lamps, so that the horofunction $\mathfrak{r}^{-,f}(g) = \mathfrak{r}^{-,l}(g)$ depends only on l, and is given by

$$\begin{aligned} &2(\max\{M(g)+1, H(g), l\} - \max\{l, 0\}) - |l - H(g)| + |l| \\ &\quad = 2\max\{M(g)+1, H(g), l\} - |l - H(g)| - l. \quad (7.6) \end{aligned}$$

Finally, the set R_l^- of negative rib horofunctions at height l has a structure similar to R_l^+.

7.4.3 Busemann Functions

Given a geodesic ray γ with $\gamma(0) = id$, let $\mathfrak{b}^{\gamma}$ denote its horofunction. Let $g \in L_2$. As discussed in Definition 7.1 and Section 7.3.1, we can define the functions m and M similarly for γ. We either have $\gamma \in \partial_\infty L_2^+$ and $m(\gamma)$ and $m(\gamma, g)$ are defined, or $\gamma \in \partial_\infty L_2^-$ and $M(\gamma)$ and $M(\gamma, g)$ are defined. The formula for $\mathfrak{b}^{\gamma}$ depends on the direction of γ, so we use $\mathfrak{b}^{+,\gamma} = \mathfrak{b}^{\gamma}$ when $\gamma \in \partial_\infty L_2^+$ and $\mathfrak{b}^{-,\gamma} = \mathfrak{b}^{\gamma}$ when $\gamma \in \partial_\infty L_2^-$, to be clear.

When $\gamma \in \partial_\infty L_2^+$, for n large enough, we apply Lemma 7.2 to obtain:

$$\begin{aligned} d(\gamma(n), id) &= 2(H(\gamma(n)) - \min\{m(\gamma), 0\}) - H(\gamma(n)), \\ d(\gamma(n), g) &= 2(H(\gamma(n)) - \min\{m(\gamma, g), H(g)\}) + H(g) - H(\gamma(n)). \end{aligned}$$

Thus the Busemann function corresponding to γ is given by

$$\mathfrak{b}^{+,\gamma}(g) = 2\left(\min\{m(\gamma), 0\} - \min\{m(\gamma, g), H(g)\}\right) + H(g). \quad (7.7)$$

If $\gamma \in \partial_\infty L_2^-$, we can similarly calculate

$$\mathfrak{b}^{-,\gamma}(g) = 2\left(\max\{M(\gamma, g) + 1, H(g)\} - \max\{M(\gamma) + 1, 0\}\right) - H(g). \tag{7.8}$$

Note that the Busemann horofunctions are obtained from the rib horofunctions by allowing the lamplighter position to approach $+\infty$ or $-\infty$ as appropriate. This is spelled out later in Observation 7.27.

Given any two horofunctions described above, one can find an element g of L_2 on which they disagree. Thus we have the following observation.

Observation 7.7 *The horofunctions $\mathfrak{s}^l$ for $l \in \mathbb{Z}$, $\pm H$, $\mathfrak{r}^{+,f}$ for $f \in L_2$ and $M(f) < H(f)$, $\mathfrak{r}^{-,f}$ for $f \in L_2$ and $m(f) \geq H(f)$, $\mathfrak{b}^{+,\gamma}$, $\mathfrak{b}^{-,\gamma}$, $\gamma \in \partial_\infty L_2$, are all pairwise distinct.*

7.5 Classification of Horofunctions

We will now prove that the functions referred to in Observation 7.7 constitute all of $\partial_h L_2$.

Definition 7.8 Given a sequence $(g_n) \subset L_2$, we say that the lamp at position k in the lamp stands of these elements *stabilizes* if there exists $N \in \mathbb{N}$ such that the lamp in position k for the lamp stand representing g_n has the same status (i.e. on or off) for all $n > N$.

We say that the lamp at position k is *flickering* if it does not stabilize.

Definition 7.9 We say that sequence (g_n) of elements of L_2 is *right stable* if there exists $N \in \mathbb{N}$ and $M \in \mathbb{Z}$ such that for all $k > M$, the lamp at position k for the lamp stand representing g_n has the same status (i.e. on or off) for all $n > N$. That is, a sequence is right stable if the set of positions of its flickering lamps (should any exist) has a maximum.

We define *left stable* similarly.

Observation 7.10 *If a sequence $(g_n) \subset L_2$ is not right stable, then there exists a subsequence (g_{n_k}) such that the sequence $(M(g_{n_k}))$ is increasing without bound.*

Similarly, if a sequence $(g_n) \subset L_2$ is not left stable, then there exists a subsequence (g_{n_k}) such that the sequence $(m(g_{n_k}))$ is decreasing without bound.

Proof If the sequence is not right stable, then $\sup\{M(g_n) \mid n \in \mathbb{N}\} = +\infty$ since if this supremum were a finite value $M_0 \in \mathbb{Z}$, then by setting

$N = 0$ and $M = M_0$, the sequence would satisfy the definition for being right stable. The existence of the desired subsequence is then guaranteed.

The proof when the sequence is not left stable is similar. □

Lemma 7.11 *Suppose that a sequence* $(g_n) \subset L_2$ *with* $H(g_n) \to l \in \mathbb{Z} \cup \{+\infty\}$ *is left stable. If* (g_n) *is associated with some horofunction* h_{g_n}*, then the set of positions of its flickering lamps (should any exist) has a minimum of at least* l*.*

Proof If (g_n) has no flickering lamps, then we are done. So assume the sequence has some flickering lamps, and let $k \in \mathbb{Z}$ be the minimum position of a flickering lamp. Suppose for contradiction that $k < l$.

Let $y \in L_2$ such that $H(y) = k$, y agrees with the stabilization of lamps of (g_n) on the positions $k-1$ and below, and the lamp at position k is off. Let $x \in L_2$ be exactly as y, except that $H(x) = k+1$. Let n be sufficiently large so that the lamps at positions $k-1$ and below of g_n have achieved their eventual status and $H(g_n) > k$.

Suppose the lamp at position k is lit in the lamp stand for g_n. In Lemma 7.2, when computing $d(g_n, x)$, $C = H(g_n) - (k+1)$, but when computing $d(g_n, y)$, $C = H(g_n) - k$, while the values for A and B remain the same (in this case, $A = k$ for both). Thus $d(g_n, x) = d(g_n, y) + 1$.

Now suppose the lamp at position k is *not* lit in the lamp stand for g_n. Using Lemma 7.2 again, when computing $d(g_n, x)$, $A = k+1$, $C = H(g_n) - (k+1)$, while when computing $d(g_n, y)$, $A = k$, $C = H(g_n) - k$, and B remains the same. In this case, we have $d(g_n, x) = d(g_n, y) - 1$.

By Equation 7.1,

$$h_{g_n}(x) - h_{g_n}(y) = \lim_{n \to \infty} d(g_n, x) - d(g_n, y)$$

which by the above, does not exist. But we assumed h_{g_n} exists. Hence, our assumption that $k < l$ is incorrect, and we have the desired result. □

Lemma 7.12 *Suppose that a sequence* $(g_n) \subset L_2$ *with* $H(g_n) \to l \in \{-\infty\} \cup \mathbb{Z}$ *is right stable. If* (g_n) *is associated with some horofunction* $h = h_{g_n}$*, then for every* $k \geq l$*, the lamp at position* k *stabilizes.*

Proof The proof for this lemma is the same as for Lemma 7.11. The asymmetry in the inequalities (one is strict, while the other is not) comes from the asymmetry of our generating set (including at but not ta). □

Lemma 7.13 *Suppose that a sequence* $(g_n) \subset L_2$ *is both left and right* *stable and that $H(g_n) \to l \in \mathbb{Z}$. If (g_n) is associated with some horofunction h_{g_n}, then there is $g \in L_2$ such that $g_n \to g$ (i.e. the sequence is eventually constant), and h_{g_n} is associated with the image of g in $\overline{L_2}$.*

Proof By Lemmas 7.11 and 7.12, all the lamps in (g_n) stabilize. Since it is stable on both sides, we in fact have the existence of some $N \in \mathbb{N}$ such that the set of lit lamps in g_n is constant for all $n > N$. Since the lamplighter limits to l by hypothesis and since $\mathbb{Z}$ is a discrete set, we have that the sequence (g_n) is eventually constant. □

Lemma 7.14 *Suppose that a sequence $(g_n) \subset L_2$ is either left or right stable,* but not both, *and that $H(g_n) \to l \in \mathbb{Z}$. If (g_n) is associated with some horofunction h_{g_n}, then h_{g_n} is a rib, i.e. one of $\mathfrak{r}^{\pm,f}$, $f \in L_2$.*

Proof We consider the case where the sequence (g_n) is left stable, but not right stable. The other case is similar.

By Lemma 7.11, there exists $N \in \mathbb{N}$ such that the the lamps below position l are stable and $H(g_n) = l$ for all $n > N$. Let $\mathfrak{r}$ be the rib horofunction that matches this stabilization. Set (r_n) to be the model sequence defined in Section 7.4.2 that generates this horofunction.

By Observation 7.10, we may take a subsequence (g_{n_k}) such that $(M(g_{n_k}))$ is increasing with $M(g_{n_k}) > k$ for all k. Choose a subsequence (r_{n_k}) of our model sequence such that $M(r_{n_k}) = M(g_{n_k})$.

Let $x \in L_2$. Choose $K \in \mathbb{N}$ such that $K > \max\{|l|, |M(x)|, |H(x)|\}$, and let $k > K$.

Let A, B, C be as in Lemma 7.2 for the computation of $d(g_{n_k}, x)$ and let A', B', C' be as in Lemma 7.2 for the computation of $d(r_{n_k}, x)$. Notice that $A = A'$ since the lamp stands for g_{n_k} and r_{n_k} are the same below the position $H(g_{n_k}) = H(r_{n_k})$, $B = M(g_{n_k}) + 1 = M(r_{n_k}) + 1 = B'$ by our choice of K, and $C = C'$ since $H(g_{n_k}) = H(r_{n_k})$. Thus, $d(g_{n_k}, x) = d(r_{n_k}, x)$.

For $x = id$, we have that $d(g_{n_k}, id) = d(r_{n_k}, id)$. Hence, $h_{g_{n_k}} = h_{r_{n_k}}$ and so therefore $h_{g_n} = \mathfrak{r}$. □

Lemma 7.15 *Suppose that a sequence $(g_n) \subset L_2$ is neither left nor right stable and that $H(g_n) \to l \in \mathbb{Z}$. If (g_n) is associated with some horofunction h_{g_n}, then $h_{g_n} = \mathfrak{s}^l$.*

Proof Suppose that there exists a subsequence (g_{n_k}) such that for all $N \in \mathbb{N}$ there exists $K_N \in \mathbb{N}$ such that for all $k > K_N$ we have that $M(g_{n_k}) > N$ and $m(g_{n_k}) < -N$. Then let $x \in L_2$, and let $N \in \mathbb{N}$ such that $N > \max\{|M(x)|, |m(x)|, |l|\}$. Let $K = \max\{K_N, K_l\}$, where

K_N is as given above and K_l is an integer such that for all $k > K_l$, $H(g_{n_k}) = l$ (recall that $H(g_n) \to l$ and the integers are a discrete set).

Let $k > K$. Then by choice of N and definition of K and using Lemma 7.2, $d(g_{n_k}, x) = 2(M(g_{n_k}) + 1 - m(g_{n_k})) - |l - H(x)|$ and specifically $d(g_{n_k}, id) = 2(M(g_{n_k}) + 1 - m(g_{n_k})) - |l|$. Thus, by Equation 7.1 $h_{g_{n_k}} = \mathfrak{s}^l(x)$, and we are done.

Now suppose that such a subsequence does *not* exist. By Observation 7.10, since (g_n) is not left stable, there exists a subsequence (g_{n_i}) such that $m(g_{n_i}) < -i$ for all i and $(m(g_{n_i}))$ is decreasing. Also by Observation 7.10, since (g_n) is not right stable, there exists a subsequence (g_{n_j}) such that $M(g_{n_j}) > j$ for all j and $(M(g_{n_j}))$ is increasing. Since these are both subsequences of (g_n), both give rise to horofunctions, and $h_{g_{n_i}} = h_{g_{n_j}} = h_{g_n}$.

Notice that the subsequence (g_{n_i}) must be right stable, otherwise we would be able to find a subsequence as in the first part of the proof. Similarly, the subsequence (g_{n_j}) must be left stable.

By Lemma 7.14, $h_{g_{n_i}}$ is equal to one of the rib examples with stable component above the lamplighter. But also by Lemma 7.14, $h_{g_{n_j}}$ is equal to one of the rib examples with stable component *below* the lamplighter. By inspecting Equations 7.3 and 7.5, we see that these two horofunctions cannot be equal, so h_{g_n} does not exist. □

Lemma 7.16 *Suppose that a sequence $(g_n) \subset L_2$ is left stable and $H(g_n) \to +\infty$. If (g_n) is associated with some horofunction h_{g_n}, then h_{g_n} is equal to a Busemann function $\mathfrak{b}^\gamma$ with $[\gamma] \in \partial_\infty L_2^+$.*

Proof By Lemma 7.11, there are no flickering lamps in (g_n), so consider the infinite lamp stand of the stabilization of lamps in (g_n). Since the sequence is left stable, if there are any lamps lit in this infinite lamp stand, there is a minimum such lamp. Thus, there exists $[\gamma] \in \partial_\infty L_2^+$ with infinite lamp stand equal to this stabilization.

Take a subsequence (g_{n_k}) such that for every positive integer K, for all $k > K$ the lamps at positions at most K in the lamp stand for g_{n_k} have achieved their eventual status and $H(g_{n_k}) > K$.

Let $x \in L_2$. Choose K large enough that $K \geq \max\{m(x), M(x), H(x)\}$, and for the finite values of $m(\gamma)$ and $m(\gamma, x)$, $K \geq \max\{m(\gamma), m(\gamma, x)\}$ as well.

Let $k > K$. Assume that h_{g_n} exists (and is therefore equal to $h_{g_{n_k}}$)

and use Lemma 7.2 and Equation 7.1:

$$h_{g_{n_k}}(x) = \lim_{n_k \to \infty} 2\Big(\max\{M(g_{n_k}, x) + 1, H(g_{n_k}), H(x)\} - \min\{m(g_{n_k}, x), H(x)\}\Big) - (H(g_{n_k}) - H(x)) - \Big[2\big(\max\{M(g_{n_k}) + 1, H(g_{n_k}), 0\} - \min\{m(g_{n_k}), 0\}\big) - H(g_{n_k})\Big].$$

Notice that if $\max\{M(g_{n_k}), M(g_{n_k}, x)\} > H(g_{n_k})$, then since $H(g_{n_k}) > M(x)$, we have that $M(g_{n_k}, x) = M(g_{n_k})$. Since $H(g_{n_k}) \geq \max\{H(x), 0\}$, we have that

$$\max\{M(g_{n_k}, x) + 1, H(g_{n_k}), H(x)\} = \max\{M(g_{n_k}) + 1, H(g_{n_k}), 0\}.$$

Therefore,

$$h_{g_{n_k}}(x) = \lim_{n_k \to \infty} 2(\min\{m(g_{n_k}), 0\} - \min\{m(g_{n_k}, x), H(x)\}) + H(x).$$

Now notice that if $m(g_{n_k}) < 0$ or $m(\gamma) < 0$, then $m(\gamma) = m(g_{n_k})$. Similarly, if $m(g_{n_k}, x) < H(x)$ or $m(\gamma, x) < H(x)$, since $H(x) < K$, then $m(g_{n_k}, x) = m(\gamma, x)$. So by Equation 7.7 and the above, $h_{g_n} = \mathfrak{b}^{\gamma}$. □

Lemma 7.17 *Suppose that a sequence $(g_n) \subset L_2$ is not left stable and $H(g_n) \to +\infty$. If (g_n) is associated with some horofunction h_{g_n}, then $h_{g_n} = H$, the height function.*

Proof By Observation 7.10, (g_n) has a subsequence (g_{n_i}) such that $(m(g_{n_i}))$ is decreasing with $m(g_{n_i}) < -i$ for all i. We still have $H(g_{n_i}) \to +\infty$, so we can further take a subsequence (g_{n_k}) such that for all k, $m(g_{n_k}) < -k$ and $H(g_{n_k}) > k$.

Let $x \in L_2$, let $K = \max\{M(x),\ |m(x)|,\ |H(x)|\}$, and consider $k > K$. By Lemma 7.2, there exists $B \in \mathbb{Z}$ such that

$$d(g_{n_k}, x) = 2(B - m(g_{n_k})) - |H(g_{n_k}) - H(x)|$$

and

$$d(g_{n_k}, id) = 2(B - m(g_{n_k})) - |H(g_{n_k})|.$$

Thus,

$$h_{g_{n_k}}(x) = \lim_{n_k \to \infty} |H(g_{n_k})| - |H(g_{n_k}) - H(x)| = H(x).$$

□

Lemma 7.18 *Suppose that a sequence $(g_n) \subset L_2$ is right stable and $H(g_n) \to -\infty$. If (g_n) is associated with some horofunction h_{g_n}, then h_{g_n} is equal to a Busemann function $\mathfrak{b}^{\gamma}$ with $[\gamma] \in \partial_\infty L_2^-$.*

Proof As in the proof of Lemma 7.16, but the Busemann function will have the lamplighter at $-\infty$ instead of $+\infty$. □

Lemma 7.19 *Suppose that a sequence $(g_n) \subset L_2$ is not right stable and $H(g_n) \to -\infty$. If (g_n) is associated with some horofunction h_{g_n}, then $h_{g_n} = -H$, the negation of the height function.*

Proof Similar to the proof of Lemma 7.17. □

Theorem 7.20 *Suppose that a sequence $(g_n) \subset L_2$ has $(H(g_n))$ converging to some value $l \in \mathbb{Z} \cup \{\pm\infty\}$. If (g_n) is associated with some horofunction h_{g_n} and $g \in L_2$, then:*

1 *if $l \in \mathbb{Z}$ and (g_n) is both left and right stable, then the sequence is eventually a constant value g_0 and h_{g_n} is in the image of L_2 in $\overline{L_2}$,*

$$h_{g_n}(g) = d(g, g_0);$$

2 *if $l = +\infty$ and (g_n) is left stable, then $h_{g_n} = \mathfrak{b}^\gamma$ for some $[\gamma] \in \partial_\infty L_2^+$,*

$$\mathfrak{b}^{+,\gamma}(g) = 2(\min\{m(\gamma), 0\} - \min\{m(\gamma, g), H(g)\}) + H(g);$$

3 *if $l = -\infty$ and (g_n) is right stable, then $h_{g_n} = \mathfrak{b}^\gamma$ for some $[\gamma] \in \partial_\infty L_2^-$,*

$$\mathfrak{b}^{-,\gamma}(g) = 2(\max\{M(\gamma, g) + 1, H(g)\} - \max\{M(\gamma) + 1, 0\}) - H(g);$$

4 *if $l \in \mathbb{Z}$ and (g_n) is neither left nor right stable, then $h_{g_n} = \mathfrak{s}^l$,*

$$\mathfrak{s}^l(g) = |l| - |l - H(g)|;$$

5 *if $l \in \mathbb{Z}$ and (g_n) is left – but not right – stable, then $h_{g_n} = \mathfrak{r}^{+,f}$ for some $f \in L_2$,*

$$\mathfrak{r}^{+,f}(g) = 2(\min\{m(f), l, 0\} - \min\{m(f, g), H(g), l\}) + \mathfrak{s}^l(g);$$

6 *if $l \in \mathbb{Z}$ and (g_n) is right – but not left – stable, then $h_{g_n} = \mathfrak{r}^{-,f}$ for some $f \in L_2$,*

$$\mathfrak{r}^{-,f}(g) = 2(\max\{M(f, g) + 1, H(g), l\} - \max\{M(f) + 1, l, 0\}) + \mathfrak{s}^l(g);$$

7 *if $l = +\infty$ and (g_n) is not left stable, then $h_{g_n} = H$;*
8 *if $l = -\infty$ and (g_n) is not right stable, then $h_{g_n} = -H$.*

Proof If $l \in \mathbb{Z}$, then apply one of Lemmas 7.13, 7.14, or 7.15, as appropriate for the existence of left or right stability. If $l = +\infty$, then apply either Lemma 7.16 or 7.17, depending on the existence of left stability. If $l = -\infty$, then apply either Lemma 7.18 or 7.19, depending on the existence of right stability. □

Lemma 7.21 *Suppose for a sequence $(g_n) \subset L_2$, $(H(g_n))$ does not converge in $\mathbb{Z} \cup \{\pm\infty\}$. Then (g_n) is* not *associated with a horofunction.*

Proof By our hypotheses, (g_n) has subsequences (g_{n_i}) and (g_{n_j}) such that $(H(g_{n_i}))$ and $(H(g_{n_j}))$ converge in $\mathbb{Z} \cup \{\pm\infty\}$, but to distinct values. By Theorem 7.20 and Observation 7.7, since these limits are distinct, $h_{g_{n_i}} \neq h_{g_{n_j}}$. Thus h_{g_n} cannot exist. □

Corollary 7.22 *Let $h \in \overline{L_2}$, and choose a sequence $(g_n) \subset L_2$ such that $h = h_{g_n}$. Then $(H(g_n))$ converges to some value $l \in \mathbb{Z} \cup \{\pm\infty\}$, and h_{g_n} can be categorized as in Theorem 7.20.*

7.6 Topology of the Horofunction Boundary

The topology of $\partial_h L_2$ is the topology of uniform convergence on compact sets. The standard basis is the collection of sets of the form

$$B_K(h, \epsilon) = \{h' \in \partial_h L_2 \mid |h(x) - h'(x)| < \epsilon \text{ for all } x \in K\},$$

where $K \subset L_2$ is compact and $\epsilon > 0$. By restricting to $0 < \epsilon < 1$, we obtain an equivalent basis. Since the minimum distance between distinct points in L_2 is 1, we may use the following sets as a basis:

$$B_K(h) = \{h' \in \partial_h L_2 \mid h(x) = h'(x) \text{ for all } x \in K\},$$

where $K \subset L_2$ is finite. Notice that pointwise convergence implies convergence in our topology since compact sets of L_2 are finite.

With the explicit descriptions of the horofunctions found in Section 7.4, we can establish the accumulation points of $\partial_h L_2$. We begin by recalling that since $\mathfrak{s}^l(g) = |l| - |l - H(g)|$, we have the following.

Observation 7.23 $\mathfrak{s}^l \to \pm H$ *as $l \to \pm\infty$.*

Observation 7.24 *The injective map that takes elements of $\partial_\infty L_2^+$ to their Busemann functions in $\partial_h L_2$ is continuous, and the same is true of $\partial_\infty L_2^-$.*

Contrast this result with Observation 7.6, which states that the injection of the union of these two sets into the horofunction boundary is not continuous. Recall that the obstruction to continuity was the non-Hausdorff property, which was proved by finding neighborhoods of distinct elements of $\partial_\infty L_2^+$ that always shared elements of $\partial_\infty L_2^-$.

Proof Let $[\gamma] \in \partial_\infty L_2^+$, and consider $B_K(\mathfrak{b}^\gamma)$ for some finite $K \subset L_2$. Let $M = \max\{M(g), H(g) \mid g \in K\}$, and let $k \in \mathbb{Z}$ such that $k > M + 2|m(\gamma)|$ if $m(\gamma) < 0$ or $k > M$ otherwise. Consider the set

$$B_{[0,k]}([\gamma], \epsilon) = \{\gamma' \in \partial_\infty L_2^+ \mid \sup\{d(\gamma(x), \gamma'(x)) \mid x \in [0,k]\} < \epsilon\}$$

for $0 < \epsilon < 1$. In [6, Observation 4.1], the authors noted that $B_{[0,k]}([\gamma], \epsilon)$ is an open set in $\partial_\infty L_2^+$. Notice that if $\gamma' \in B_{[0,k]}([\gamma], \epsilon)$, then the lamp stands of γ and γ' agree on all lamps at positions M or below. Thus, by Equation (7.7), $\mathfrak{b}^\gamma(g) = \mathfrak{b}^{\gamma'}(g)$ for all $g \in K$. Therefore, $\mathfrak{b}^{\gamma'} \in B_K(\mathfrak{b}^\gamma)$ for all $\gamma' \in B_{[0,k]}([\gamma], \epsilon)$, and so our injection is continuous.

The proof for the injection of $\partial_\infty L_2^-$ is similar. □

The topology of each of these sets is a punctured Cantor set, but in $\partial_h L_2$ these punctures are "filled" by the height function and its negative, as we now show.

Observation 7.25 *If $([\gamma_n]) \subset \partial_\infty L_2^+$ with $m(\gamma_n) \to -\infty$, then $\mathfrak{b}^{\gamma_n} \to H$. Similarly, if $([\gamma_n]) \subset \partial_\infty L_2^-$ with $M(\gamma_n) \to +\infty$, then $\mathfrak{b}^{\gamma_n} \to -H$.*

Proof Let $([\gamma_n]) \subset \partial_\infty L_2^+$ with $\lim m(\gamma_n) = -\infty$. By Equation (7.7),

$$\mathfrak{b}^{\gamma_n}(g) = 2\left(\min\{m(\gamma_n), 0\} - \min\{m(\gamma_n, g), H(g)\}\right) + H(g).$$

Fix g and take n large enough so that $m(\gamma_n) < \min\{0, m(g), H(g)\}$, then

$$\mathfrak{b}^{\gamma_n}(g) = 2\left(m(\gamma_n) - m(\gamma_n)\right) + H(g) = H(g).$$

Thus, $\mathfrak{b}^{\gamma_n} \to H$. The other proof is similar. □

For a given $l \in \mathbb{Z}$, the following observation remarks that the spine is an accumulation point of the positive and negative rib functions. The proofs are calculations similar to those in Observation 7.25.

Observation 7.26 *Let $(f_n^l) \subset L_2$ be a sequence satisfying $M(f_n^l) < H(f_n^l) = l$ and $m(f_n^l) \to -\infty$ as $n \to \infty$. Then $\mathfrak{r}^{+,f_n^l} \to \mathfrak{s}^l$.*

Similarly, if $(f_n^l) \subset L_2$ is a sequence satisfying $m(f_n^l) \geq H(f_n^l) = l$ and $M(f_n^l) \to \infty$ as $n \to \infty$, then $\mathfrak{r}^{-,f_n^l} \to \mathfrak{s}^l$.

Finally, the ribs accumulate to Busemann functions.

Observation 7.27 *For a geodesic ray γ, with $\gamma(0) = id$, set $f_n = \gamma(n)$. If $\gamma \in \partial_\infty L_2^+$, then for large enough n, each f_n defines $\mathfrak{r}^{+,f_n}$ and $\mathfrak{r}^{+,f_n} \to \mathfrak{b}^{+,\gamma}$. If $\gamma \in \partial_\infty L_2^-$, then for large enough n each f_n defines $\mathfrak{r}^{-,f_n}$ and $\mathfrak{r}^{+,f_n} \to \mathfrak{b}^{-,\gamma}$.*

Proof We consider the $\gamma \in \partial_\infty L_2^+$ case. For large enough n, each f_n satisfies the requirements for defining $\mathfrak{r}^{+,f_n}$. Let $g \in L_2$ be given, and consider Equations (7.3) and (7.7). Again for large enough n, $m(f_n) = m(\gamma)$ and $m(f_n, g) = m(\gamma, g)$. Thus

$$\mathfrak{r}^{+,f_n} - \mathfrak{b}^{+,\gamma} = \mathfrak{s}^{H(f_n)}(g) - H(g) \to 0 \text{ as } n \to \infty.$$

□

With Observations 7.23, 7.25, 7.26, and 7.27, we have the picture of the horofunction boundary illustrated in Figure 7.1 in the introduction.

7.7 Action of L_2 on the Horofunction Boundary

We now conclude with a few comments about the action of L_2 on $\partial_h L_2$.

An isometric action of a group G on a metric space (X, d) with base point b can be extended to the horofunction boundary $\partial_h X$ in the following way. For $g \in G$ and $(y_n) \subset X$ giving rise to a horofunction, we have that

$$g \cdot h_{y_n}(x) = h_{g \cdot y_n}(x) = \lim_{n \to \infty} d(g \cdot y_n, x) - d(g \cdot y_n, b).$$

In our setting, the action of L_2 on itself is by left multiplication. We compose lamp stands $g_1 \cdot g_2$ by starting with the lamp stand for g_1 and having the lamplighter move and toggle lamps as in g_2, but from a starting position of $H(g_1)$ rather than 0.

Observation 7.28 *Let $g \in L_2$, $h \in \overline{L_2}$, and choose $(g_n) \subset L_2$ such that $h = h_{g_n}$. Then $H(g \cdot g_n) \to H(g) + \lim H(g_n)$, where for $k \in \mathbb{Z}$, $\pm\infty + k$ is understood to mean $\pm\infty$. Also, $(g \cdot g_n)$ is left (resp. right) stable, if and only if (g_n) is left (resp. right) stable.*

Proof These statements all follow from the fact that the lamp stand for g has only finitely many lit lamps and the lamplighter is at a finite position. □

Corollary 7.29 *Each of the categories of horofunctions in $\overline{L_2}$ described in Theorem 7.20 is invariant under the action of L_2.*

Proof This result follows from Observation 7.28 and Corollary 7.22. □

Interestingly, this implies the following.

Corollary 7.30 *The height function H and its negation are global fixed points of the action of L_2 on $\partial_h L_2$.*

We now consider the action of L_2 on each of the other categories of horofunctions.

Let $g \in L_2$. The action of g on $\partial_\infty L_2$ is described in [6, Sections 3.4 and 4.6]. If $H(g) \neq 0$, then the action of g on $\partial_\infty L_2$ has two fixed points, which are given the notation g^∞ and $g^{-\infty}$ in [6]. If $H(g) > 0$, then $g^\infty \in \partial_\infty L_2^+$ and $g^{-\infty} \in \partial_\infty L_2^-$. Otherwise, the reverse is true. In the topology of $\partial_\infty L_2$, the action of g has north-south dynamics with attractor g^∞ and repeller $g^{-\infty}$. Recall that in $\partial_\infty L_2$, the punctures in the two Cantor sets are "filled" by points from the opposite Cantor set, while in $\partial_h L_2$, these punctures are filled by H and $-H$. Thus, in the horofunction boundary we see similar dynamics with the visual boundary, except it occurs on the separate sets of $\partial_\infty L_2^+ \cup \{H\}$ and $\partial_\infty L_2^- \cup \{-H\}$.

Observation 7.31 *For $g \in L_2$ with $H(g) \neq 0$, the action of g on $\partial_h L_2$ has four fixed points: $H, -H, \mathfrak{b}^{g^\infty}, \mathfrak{b}^{g^{-\infty}}$. The action of g has north-south dynamics on $\partial_\infty L_2^+ \cup \{H\}$ with poles H and either g^∞ or $g^{-\infty}$ (whichever is in the set) and also on $\partial_\infty L_2^- \cup \{-H\}$ with poles $-H$ and either g^∞ or $g^{-\infty}$ (whichever is in the set). The point g^∞ is always an attractor and the point $g^{-\infty}$ is always a repeller. If $H(g) > 0$, then H is an attractor and $-H$ is a repeller. If $H(g) < 0$, then these roles are reversed.*

For a spinal horofunction $\mathfrak{s}^l \in \partial_h L_2$, $l \in \mathbb{Z}$, the action of g on $\mathfrak{s}^l$ is given by $g \cdot \mathfrak{s}^l = \mathfrak{s}^{H(g)+l}$.

We see similar behavior on the ribs of $\partial_h L_2$ in that the l value is translated by the height of the group element, but there is also additional structure in this case. Let $g \in L_2$ and let $f \in L_2$ such that $\mathfrak{r}^{+,f}$ exists (i.e. $M(f) < H(f)$). Then $g \cdot \mathfrak{r}^{+,f} = \mathfrak{r}^{+,\overline{gf}}$, where $\overline{gf}$ has the lamp stand for gf but with all of the lamps at position $H(gf) = H(g) + H(f)$ and above switched off. Note that g acts as a bijection from $R^+_{H(f)}$ to $R^+_{H(g)+H(f)}$.

Notice that if $m(g) \neq H(g) + m(f)$, then

$$m(\overline{gf}) = \min\{H(g) + m(f), m(g)\}.$$

Using the notation in Section 7.4.2, the above yields the following description of the action on rib horofunctions that are "close" to the spine.

Observation 7.32 *Let $g \in L_2$ and $l \in \mathbb{Z}$. Let $k < l$ such that $H(g) + k < m(g)$. The action of g on $\partial_h L_2$ restricted to the subset $\hat{m}_l^{-1}(k)$ of R_l^+ is a bijection onto the subset $\hat{m}^{-1}_{H(g)+l}(H(g) + k)$ of $R^+_{H(g)+l}$.*

Corollary 7.33 *Let $g \in L_2$ such that $H(g) = 0$ and let $l \in \mathbb{Z}$. If*

$k < \min\{m(g) - H(g), l\}$, then the subset $\hat{m}_l^{-1}(k)$ of R_l^+ is invariant under the action of g.

The action on such a rib $\mathfrak{r}^{+,f}$ leaves $m(f)$ and $H(f)$ fixed, but changes the status of lamps between those positions. This gives a permutation on the set $m_l^{-1}(k)$.

The similar statements also hold for negative ribs.

Acknowledgments

The authors thank Moon Duchin for helpful conversations and the anonymous referee for many thoughtful suggestions. Support for the first author from a summer research grant from SUNY Oneonta is gratefully acknowledged. Both authors also thank the Institute for Advanced Study for its hospitality at the Park City Math Institute Summer Session 2012.

References

[1] Bridson, Martin R. and Haefliger, André. 1999. *Metric spaces of non-positive curvature.* Grundlehren der Mathematischen Wissenschaften [Fundamental Principles of Mathematical Sciences], vol. 319. Springer-Verlag.

[2] Develin, Mike. 2002. Cayley compactifications of abelian groups. *Ann. Comb.*, **6**(3-4), 295–312.

[3] Friedland, Shmuel and Freitas, Pedro J. 2004. p-metrics on $\mathrm{GL}(n, \mathbb{C})/U_n$ and their Busemann compactifications. *Linear Algebra Appl.*, **376**, 1–18.

[4] Grigorchuk, Rostislav I. and Żuk, Andrzej. 2001. The lamplighter group as a group generated by a 2-state automaton, and its spectrum. *Geom. Dedicata*, **87**(1-3), 209–244.

[5] Gromov, M. 1981. Hyperbolic manifolds, groups and actions. Pages 183–213 of: *Riemann surfaces and related topics: Proceedings of the 1978 Stony Brook Conference (State Univ. New York, Stony Brook, N.Y., 1978).* Ann. of Math. Stud., vol. 97. Princeton University Press.

[6] Jones, Keith and Kelsey, Gregory A. 2015. Visual boundaries of Diestel-Leader graphs. *Topology Proc.*, **46**, 181–204.

[7] Kitzmiller, Kyle and Rathbun, Matt. 2011. The visual boundary of $\mathbb{Z}^2$. *Involve*, **4**(2), 103–116.

[8] Klein, Tom and Nicas, Andrew. 2009. The horofunction boundary of the Heisenberg group. *Pacific J. Math.*, **242**(2), 299–310.

[9] Klein, Tom and Nicas, Andrew. 2010. The horofunction boundary of the Heisenberg group: the Carnot-Carathéodory metric. *Conform. Geom. Dyn.*, **14**, 269–295.

[10] Rieffel, Marc A. 2002. Group C^*-algebras as compact quantum metric spaces. *Doc. Math.*, **7**, 605–651.

[11] Stein, Melanie and Taback, Jennifer. 2013. Metric properties of Diestel-Leader groups. *Michigan Math. J.*, **62**(2), 365–386.

[12] Walsh, Cormac. 2009. Busemann points of Artin groups of dihedral type. *Internat. J. Algebra Comput.*, **19**(7), 891–910.

[13] Walsh, Cormac. 2011. The action of a nilpotent group on its horofunction boundary has finite orbits. *Groups Geom. Dyn.*, **5**(1), 189–206.

[14] Webster, Corran and Winchester, Adam. 2005. Boundaries of hyperbolic metric spaces. *Pacific J. Math.*, **221**(1), 147–158.

[15] Webster, Corran and Winchester, Adam. 2006. Busemann points of infinite graphs. *Trans. Amer. Math. Soc.*, **358**(9), 4209–4224.

[16] Woess, Wolfgang. 2005. Lamplighters, Diestel-Leader graphs, random walks, and harmonic functions. *Combin. Probab. Comput.*, **14**(3), 415–433.

8

Intrinsic Geometry of a Euclidean Simplex

Barry Minemyer

Abstract

We give a simple technique to compute the distance between two points in an n-dimensional Euclidean simplex, where the points are given in barycentric coordinates, using only the edge lengths of that simplex. We then use this technique to verify a few computations which will be used in subsequent papers. The most important application is a formula for intrinsically computing the volume of a Euclidean simplex which is more efficient (and more natural) than any previously documented methods.

8.1 Introduction

While studying the isometric embedding problem for metric simplicial complexes in [9], the author came across the following basic problem. In attempting to work out basic examples, one needs to be able to compute distances between points in a given Euclidean simplex given only the barycentric coordinates of those points and the lengths of the edges of that simplex. More specifically, let $\sigma = \langle v_0, v_1, ..., v_n \rangle$ be an (abstract) n-dimensional simplex with vertices $v_0, ..., v_n$, let e_{ij} denote the edge connecting the vertices v_i and v_j, let γ_{ij} denote the length assigned to e_{ij}, and let $x, y \in \sigma$ with barycentric coordinates $x = (x_0, ..., x_n)$ and $y = (y_0, ..., y_n)$. The questions that needed to be answered were:

1 for what values of γ_{ij} can σ be realized as a legitimate simplex in $\mathbb{E}^n$;

2 assuming that we have "good" values γ_{ij}, give a simple formula to calculate $d_\sigma(x, y)$, the length of the straight line segment connecting x to y within σ.

A naive attempt to solve question (2) is to construct an explicit isometric embedding of σ into $\mathbb{E}^n$, and then compute $d_\sigma(x,y)$ using basic Euclidean geometry. But for $n \geq 3$ constructing this embedding becomes quite cumbersome, and a much simpler method is described in Section 8.2.

Question (1) is an old problem, and was first solved by Cayley in [3] way back in 1841. Question (1) was also solved for hyperbolic and spherical simplices by Karliğa in [5]. Another solution to question (1), as well as the main ingredient to the solution to question (2), can be found in an arXiv paper by Igor Rivin [11]. But Rivin does not use his formula for the Gram matrix to show how to compute distances interior to a simplex, which is the main issue that we take up in this chapter. In Sections 8.3, 8.4, and 8.5 we demonstrate the power of formula (8.3) by working out some computations which would be difficult to produce directly. Most notably though is Theorem 8.3 which, in conjunction with equation (8.3), gives a simple formula for intrinsically computing the volume of any given Euclidean n-simplex. This formula is computationally simpler than the widely used *Cayley–Menger determinant*, as will be discussed in Section 8.6.

8.2 The Main Formula

Linearly embed the n-simplex σ into $\mathbb{R}^n$ in some way, and by abuse of notation we will identify each vertex v_i with its image in $\mathbb{R}^n$. For each i let $w_i := v_i - v_0$, so w_i is just the vector in $\mathbb{R}^n$ representing the edge e_{0i}. Since σ is embedded in $\mathbb{R}^n$, the collection $\{w_i\}_{i=1}^n$ forms a basis for $\mathbb{R}^n$. If we had values for $\langle w_i, w_j\rangle$ then we could use those values to define a symmetric bilinear form on $\mathbb{R}^n$. But observe that due to the symmetry and bilinearity of $\langle , \rangle$:

$$\langle w_i - w_j, w_i - w_j\rangle = \langle w_i, w_i\rangle - 2\langle w_i, w_j\rangle + \langle w_j, w_j\rangle$$

and so

$$\langle w_i, w_j\rangle = \frac{1}{2}\left(\langle w_i, w_i\rangle + \langle w_j, w_j\rangle - \langle w_i - w_j, w_i - w_j\rangle\right). \tag{8.1}$$

Now, if our embedding of σ into $\mathbb{R}^n$ were an isometry, then for all i we would have $\langle w_i, w_i\rangle = \gamma_{0i}^2$. Equation (8.1) would then become

$$\langle w_i, w_j\rangle = \frac{1}{2}(\gamma_{0i}^2 + \gamma_{0j}^2 - \gamma_{ij}^2). \tag{8.2}$$

where $\gamma_{ij}^2 := 0$ if $i = j$.

The trick now is to *define* the symmetric bilinear form $\langle , \rangle$ by equation (8.2). This naturally defines a quadratic form Q on $\mathbb{R}^n$ (using only the edge lengths assigned to σ, and our original choice of embedding). It is easy to see that an orthogonal transformation will map our original embedding to an isometric embedding with respect to the standard Euclidean metric if and only if this form Q is positive definite. A simple proof can be found in the first few pages of [2], and, again, the above result can also be found in [11]. This completes the solution to question (1).

To solve question (2), let $x, y \in \sigma$ with barycentric coordinates $(x_i)_{i=0}^n$ and $(y_i)_{i=0}^n$, respectively. Just as before we consider some linear embedding of σ into $\mathbb{R}^n$ and abuse notation by associating x and y with their images in $\mathbb{R}^n$. Note then that $x = \sum_{i=0}^n x_i v_i$ and $y = \sum_{i=1}^n y_i v_i$. The square of the distance $d_\sigma(x, y)$ between x and y in σ is given by $\langle x - y, x - y \rangle$, where $\langle , \rangle$ is the symmetric bilinear form defined above. What is left to do is to show how to use equation (8.2) to produce a nice formula to compute $d_\sigma(x, y)$.

Define the quadratic form Q as above and note that, with respect to our basis $\{w_i\}_{i=1}^n$, we can express Q as an $(n \times n)$ symmetric matrix by

$$Q_{ij} = (\langle w_i, w_j \rangle)_{ij} = \left(\frac{1}{2} \left(\gamma_{0i}^2 + \gamma_{0j}^2 - \gamma_{ij}^2 \right) \right)_{ij} . \tag{8.3}$$

Recall that, by the definition of barycentric coordinates,

$$x_0 = 1 - \sum_{i=1}^n x_i \qquad \text{and} \qquad y_0 = 1 - \sum_{i=1}^n y_i. \tag{8.4}$$

With the help of equation (8.4) we compute

$$\begin{aligned} x - y &= \sum_{i=0}^n (x_i - y_i) v_i = (x_0 - y_0) v_0 + \sum_{i=1}^n (x_i - y_i) v_i \\ &= - \sum_{i=1}^n (x_i - y_i) v_0 + \sum_{i=1}^n (x_i - y_i) v_i = \sum_{i=1}^n (x_i - y_i) w_i. \end{aligned} \tag{8.5}$$

Combining equations (8.3) and (8.5) yields

$$\langle x - y, x - y \rangle = \left\langle \sum_{i=1}^n (x_i - y_i) w_i, \sum_{j=1}^n (x_j - y_j) w_j \right\rangle$$

$$= \sum_{i,j=1}^{n} (x_i - y_i)(x_j - y_j)\langle w_i, w_j \rangle = [x-y] \cdot Q[x-y], \qquad (8.6)$$

where "$\cdot$" represents the standard Euclidean inner product, and where $[x-y]$ is the vector in $\mathbb{R}^n$ defined by $[x-y] = (x_i - y_i)_{i=1}^n$. Both the matrix Q and the vector $[x-y]$ are expressed using only the barycentric coordinates of x and y and the edge lengths assigned to σ. So, using equation (8.6), computing distances in σ requires only matrix multiplication.

8.3 The Minimal Allowable Edge Length when all Other Edges have Length 1

Let σ be as above, and assume that all edges of σ have length 1 except one edge whose length we will denote by α. By symmetry, let e_{0n} be the edge with length α, i.e. $\gamma_{0n} = \alpha$. The question is, for what values of α does σ admit an affine isometric embedding into $\mathbb{R}^n$? When $n = 2$ it is easy to see that $0 < \alpha < 2$, and for $n = 3$ one observes that $0 < \alpha < \sqrt{3}$. But it starts to get a little more subtle[1] once $n \geq 4$. Note that the quadratic form Q from Section 8.2 is

$$Q(\alpha) = \begin{bmatrix} 1 & \frac{1}{2} & \cdots & \frac{1}{2} & \frac{1}{2}\alpha^2 \\ \frac{1}{2} & 1 & \cdots & \frac{1}{2} & \frac{1}{2}\alpha^2 \\ \vdots & \vdots & \ddots & \vdots & \vdots \\ \frac{1}{2} & \frac{1}{2} & \cdots & 1 & \frac{1}{2}\alpha^2 \\ \frac{1}{2}\alpha^2 & \frac{1}{2}\alpha^2 & \cdots & \frac{1}{2}\alpha^2 & \alpha^2 \end{bmatrix}$$

We need to find the values of α for which $Q(\alpha)$ is positive definite. We first need a Lemma.

Lemma 8.1 *Let A_n and B_n denote the $n \times n$ matrices*

$$A_n = \begin{bmatrix} 1 & \frac{1}{2} & \cdots & \frac{1}{2} \\ \frac{1}{2} & 1 & \cdots & \frac{1}{2} \\ \vdots & \vdots & \ddots & \vdots \\ \frac{1}{2} & \frac{1}{2} & \cdots & 1 \end{bmatrix}, \qquad B_n = \begin{bmatrix} \frac{1}{2} & \frac{1}{2} & \cdots & \frac{1}{2} \\ 1 & \frac{1}{2} & \cdots & \frac{1}{2} \\ \vdots & \ddots & \cdots & \vdots \\ \frac{1}{2} & \cdots & 1 & \frac{1}{2} \end{bmatrix}.$$

[1] When doing some research for [10], I once assumed $\alpha = \frac{3}{2}$ would always work. To my surprise I found out that this leads to a degenerate simplex when $n = 9$, and does not lead to a realizable Euclidean simplex for all larger dimensions.

Then $det(A_n) = \frac{n+1}{2^n}$ *and* $det(B_n) = \frac{(-1)^{n+1}}{2^n}$.

Proof The proof proceeds by (simultaneous) induction on n. The base cases are easily checked and left to the reader.

Let us first compute $\det(A_n)$. Add the negative of the first row to the nth row, which does not change the determinant. Then cofactor expansion along the new n^{th} row, along with using both portions of the induction hypothesis, yields

$$\begin{aligned}\det(A_n) &= (-1)^{n+1}\left(\frac{-1}{2}\right)\det(B_{n-1}) + \frac{1}{2}\det(A_{n-1}) \\ &= (-1)^{n+1}\left(\frac{-1}{2}\right)\frac{(-1)^n}{2^{n-1}} + \left(\frac{1}{2}\cdot\frac{n}{2^{n-1}}\right) \\ &= \frac{1}{2^n} + \frac{n}{2^n} = \frac{n+1}{2^n}.\end{aligned}$$

To compute $\det(B_n)$, add the negative of row 1 to row n. The only term in the new n^{th} row which is not 0 is the second to last term, and it is $\frac{1}{2}$. Then cofactor expansion along the last row gives

$$\det(B_n) = -\frac{1}{2}\det(B_{n-1}) = -\frac{1}{2}\frac{(-1)^n}{2^{n-1}} = \frac{(-1)^{n+1}}{2^n}.$$

□

With the aid of Lemma 8.1 we are now prepared to prove the following Theorem.

Theorem 8.2 *The quadratic form* $Q(\alpha)$ *is positive definite if and only if* $0 < \alpha < \sqrt{\frac{2n}{n-1}}$.

Remark Note that $\sqrt{\frac{2n}{n-1}}$ is a decreasing function, and $\lim_{n\to\infty}\sqrt{\frac{2n}{n-1}} = \sqrt{2}$. So all values of α with $0 < \alpha < \sqrt{2}$ always lead to a Euclidean simplex. This fact is used in [10].

Proof of Theorem 8.2 By Lemma 8.1 all of the minors of $Q(\alpha)$ which contain the $(1,1)$ entry are positive. Thus, $Q(\alpha)$ is positive definite if and only if $\det(Q) > 0$. To compute $\det(Q)$, factor an α^2 out of both the nth column and the nth row. This yields

$$\det(Q) = \alpha^4 \begin{vmatrix} 1 & \frac{1}{2} & \cdots & \frac{1}{2} & \frac{1}{2} \\ \frac{1}{2} & 1 & \cdots & \frac{1}{2} & \frac{1}{2} \\ \vdots & \vdots & \ddots & \vdots & \vdots \\ \frac{1}{2} & \frac{1}{2} & \cdots & 1 & \frac{1}{2} \\ \frac{1}{2} & \frac{1}{2} & \cdots & \frac{1}{2} & \frac{1}{\alpha^2} \end{vmatrix}.$$

Note that we are assuming that $\alpha > 0$ since it is the side length of a non-degenerate simplex.

As in the proof of Lemma 8.1 we add the negative of the first row to the nth row and then use cofactor expansion along the nth row to obtain

$$\begin{aligned} \det(Q) &= \alpha^4 \left((-1)^{n+1} \left(-\frac{1}{2} \right) \det(B_{n-1}) + \left(\frac{1}{\alpha^2} - \frac{1}{2} \right) \det(A_{n-1}) \right) \\ &= \alpha^4 \left((-1)^n \left(\frac{1}{2} \right) \frac{(-1)^n}{2^{n-1}} + \left(\frac{1}{\alpha^2} - \frac{1}{2} \right) \frac{n}{2^{n-1}} \right) \\ &= \alpha^2 \left(\frac{\alpha^2(1-n)}{2^n} + \frac{n}{2^{n-1}} \right), \end{aligned} \tag{8.7}$$

where the notation A_n and B_n comes from Lemma 8.1. We then see from equation (8.7) that

$$\begin{aligned} \det(Q) > 0 \quad &\Longleftrightarrow \quad \frac{\alpha^2(1-n)}{2^n} + \frac{n}{2^{n-1}} > 0 \\ &\Longleftrightarrow \quad \alpha < \sqrt{\frac{2n}{n-1}}. \end{aligned} \tag{8.8}$$

□

8.4 Volume of an n-simplex via Edge Lengths

Theorem 8.3 *Let $\sigma = \langle v_0, v_1, \ldots, v_n \rangle$ be an n-dimensional Euclidean simplex with edge lengths $\{\gamma_{ij}\}_{i,j=0}^n$. Let Q be the $n \times n$ matrix defined by*

$$Q_{ij} = \left(\frac{1}{2} \left(\gamma_{0i}^2 + \gamma_{0j}^2 - \gamma_{ij}^2 \right) \right).$$

Then

$$Vol(\sigma) = \frac{1}{n!} \sqrt{det(Q)} \tag{8.9}$$

Proof Let σ, $\langle v_i \rangle_{i=0}^n$, $\{\gamma_{ij}\}_{i,j=1}^n$, and Q be as above. Isometrically embed σ into $\mathbb{E}^n$, and let $w_i := v_i - v_0$ for all i. Let $W = [w_1\, w_2 \ldots w_n]$ be the $n \times n$ matrix whose columns are the vectors in $\{w_i\}_{i=1}^n$. It is well known that $\det(W)$ is the volume of the parallelpiped spanned by the vectors in $\{w_i\}_{i=1}^n$. Thus $\mathrm{Vol}(\sigma) = \frac{1}{n!}\det(W)$. But notice that $W^T W = (w_i \cdot w_j)_{ij} = Q$. So $\det(Q) = \det(W)^2$, which proves the theorem. □

Remark Combining Theorem 8.3 with equation (8.3) produces a very nice formula for *intrinsically* computing the volume of an n-simplex, meaning that the formula depends only on the assigned edge lengths and not on the coordinates of any of the vertices. The current technique for finding such volumes is by using the *Cayley–Menger determinant.* This will be discussed in Section 8.6. For now, it is worth pointing out that computing the same volume using a Cayley–Menger determinant involves computing an $(n+2) \times (n+2)$ determinant and, in the author's opinion, is much less natural.

Comment The only formulas that I am aware of for computing the volume of an n-simplex either require coordinates for the vertices of the simplex or are more difficult computationally[2] (and much less natural) than the formula given here.

In [6] we are interested in knowing the edge length of an equilateral n-simplex whose volume is 1. To compute this, let e_n denote the common edge length of σ. Then $Q = (e_n)^2 A_n$, where A_n is the notation used in Lemma 8.1. Hence

$$1 = \mathrm{Vol}(\sigma) = \frac{1}{n!}\sqrt{\det(Q)} = \frac{(e_n)^n}{n!}\sqrt{\frac{n+1}{2^n}}.$$

Solving for e_n yields

$$e_n = \left(n! \sqrt{\frac{2^n}{n+1}} \right)^{\frac{1}{n}}. \tag{8.10}$$

As a last note, it is interesting to consider $\lim_{n\to\infty} e_n$. It is not hard to check that

$$\lim_{n\to\infty} (n!)^{\frac{1}{n}} = \infty \qquad \text{and} \qquad \lim_{n\to\infty} \left(\sqrt{\frac{2^n}{n+1}} \right)^{\frac{1}{n}} = \sqrt{2}$$

and thus $\lim_{n\to\infty} e_n = \infty$. So, for an equilateral n-simplex to have volume

[2] Our formula here requires computing the determinant of an $n \times n$ matrix. Other formulas require determinants of $(n+1) \times (n+1)$ matrices or larger.

1, as n approaches infinity the edge lengths must approach infinity as well. The geometric intuition here is to notice that the equilateral n-simplex with unit edge lengths consists of a smaller percentage of the unit hypercube as n gets larger. Thus, the volume of the simplex decreases in n, and so the edge lengths must increase to make the volume 1.

8.5 Distance from the Barycenter to the Boundary of an Equilateral Simplex

As a final example of the utility of our matrix Q, let us compute the distance from the barycenter to the boundary of an equilateral simplex. Let σ be an equilateral simplex, so that all edge lengths are the same length $\gamma := \gamma_{ij}$. Let b denote the barycenter of σ, meaning that b has barycentric coordinates $\left(\frac{1}{n+1}\right)_{i=1}^{n}$. Since σ is equilateral, the distance from b to the boundary of σ is equal to the distance from b to the barycenter of any of the codimension 1 faces of σ. For convenience, we compute the distance from b to b', where b' is the barycenter of the codimension 1 face opposite the vertex v_0. So b' has barycentric coordinates $\left(0, \frac{1}{n}, \frac{1}{n}, \ldots, \frac{1}{n}\right)$.

Observe that

$$[b' - b] = \left(\frac{1}{n(n+1)}\right)_{i=1}^{n} \qquad \text{and} \qquad Q_{ij} = \begin{cases} \gamma^2 & \text{if } i = j, \\ \frac{1}{2}\gamma^2 & \text{if } i \neq j. \end{cases}$$

A simple calculation then shows that

$$[b' - b] \cdot Q[b' - b] = \frac{\gamma^2}{n^2(n+1)^2}\left(\frac{n(n+1)}{2}\right) = \frac{\gamma^2}{2n(n+1)}.$$

Thus,

$$d_\sigma(b, \partial\sigma) = \frac{\gamma}{\sqrt{2n(n+1)}}. \tag{8.11}$$

Of particular interest is knowing $d_\sigma(b, \partial\sigma)$ when σ is an equilateral simplex with volume 1. Using the notation from Section 8.4 we have that $\gamma = e_n$, and combining equations (8.10) and (8.11) yields

$$d_\sigma(b, \partial\sigma) = \frac{e_n}{\sqrt{2n(n+1)}} = \left(n!\sqrt{\frac{2^n}{n+1}}\right)^{\frac{1}{n}} \cdot \left(\frac{1}{\sqrt{2n(n+1)}}\right)$$
$$= \left(\frac{n!}{\sqrt{n^n(n+1)^{n+1}}}\right)^{\frac{1}{n}}.$$

8.6 Cayley–Menger Determinants and Gromov's $\mathcal{K}$-curvature Question

Given an oriented n-simplex $\sigma = \langle v_0, v_1, \dots, v_n\rangle$ with associated edge lengths $\{\gamma_{ij}\}$, one could organize this data into an $(n+1) \times (n+1)$ matrix $B = (b_{ij})$ defined by

$$b_{ij} = \gamma_{ij}^2, \tag{8.12}$$

where $\gamma_{ii} := 0$.

The first thing to point out is that the matrix Q defined by equation (8.3) and the matrix B in equation (8.12) are "equivalent" in the sense that they are uniquely determined by the exact same data. Moreover, this data (the edge lengths) can be easily recovered from either matrix. Therefore, given Q it is easy to construct B, and vice versa.

The matrix B mentioned above was first considered by Cayley in [3], and independently studied 80 years later by Menger in [8]. They used the matrix B to intrinsically compute the volume of σ just as in Theorem 8.3. This volume formula is:

$$\mathrm{Vol}(\sigma) = \left(\frac{(-1)^{n-1}}{2^n(n!)^2}\det(\bar{B})\right)^{\frac{1}{2}}, \tag{8.13}$$

where $\bar{B}$ is the $(n+2) \times (n+2)$ matrix obtained by placing B in the bottom right hand corner, adding a top row of $(0, 1, 1, \dots, 1)$, and a left column of $(0, 1, 1, \dots, 1)^T$. The determinant $\det(\bar{B})$ is often referred to as the *Cayley–Menger determinant*.

As the matrices Q and B are defined using the exact same data, the formulas (8.9) and (8.13) have some similarities. But formula (8.9) is certainly computationally simpler because it requires computing an $n \times n$ determinant instead of an $(n+2) \times (n+2)$ determinant. Of course, one makes a slight sacrifice in the simplicity of the matrix representation when considering Q over B. But in return one gains a simpler volume formula as well as a natural means of computing distances and other geometric quantities in an intrinsic manner.

In closing, we relate the matrix Q to Gromov's $\mathcal{K}$-curvature problem found in [4]. The following is taken almost directly from [4].

Let $(\mathcal{X}, d)$ be an arbitrary metric space, and let M_r denote the space of positive symmetric $r \times r$ matrices. Let $K_r(\mathcal{X}) \subset M_r$ denote the subset realizable by the distances among r-tuples of points as given by the matrix B in equation (8.12). i.e., the $r \times r$ matrix[3] $B = (b_{ij}) \in K_r(\mathcal{X})$ if and only if there exists an r-tuple of points $(x_1, \ldots, x_r) \in \mathcal{X}^r$ such that $d(x_i, x_j)^2 = b_{ij}$ for all i, j. Then every subset $\mathcal{K} \subset M_r$ defines the *global $\mathcal{K}$-curvature class*, which consists of the spaces $\mathcal{X}$ with $K_r(\mathcal{X}) \subset \mathcal{K}$, and the *local $\mathcal{K}$-curvature class*, where each point $x \in \mathcal{X}$ is required to admit a neighborhood U with $K_r(U) \subset \mathcal{K}$. Gromov's curvature problem is then as follows

Gromov's Curvature Problem. Given $\mathcal{K} \subset M_r$, describe the spaces $\mathcal{X}$ in the $\mathcal{K}$-curvature class.

This problem was answered in some very specific casess by Gromov in [4]. When $r = 4$, this problem was solved (in the global setting) by Berg and Nikolaev in [1] for CAT(0) spaces, and by Lebedeva and Petrunin in [7] for spaces whose curvature is bounded below. But all of these solutions deal with the data in the matrix B and not the actual matrix B itself.

In light of the discussion above, we can equivalently replace the matrix B with the matrix Q when discussing Gromov's curvature problem. But the matrix Q seems to more closely capture the geometry of the points involved. So one could ask the same question but look for answers which intrinsically depend on Q instead of inequalities using the specific distances. For example, one could ask how knowledge of the eigenvalues and associated eigenspaces of such matrices Q affect the geometry of the underlying space $\mathcal{X}$, and vice versa.

Acknowledgements

Ross Geoghegan has had a tremendous positive influence on the author's career, and as such the author is honored to have a chapter published in these conference proceedings. The author would also like to thank the

[3] Technically, Gromov considers $b_{ij} = d(x_i, x_j)$ instead of $b_{ij} = d(x_i, x_j)^2$. But these two formulations are clearly equivalent.

anonymous referee both for catching various typographical errors and for some nice mathematical suggestions.

References

[1] Berg, I. D. and Nikolaev, I. G. 2008. Quasilinearization and curvature of Aleksandrov spaces. *Geom. Dedicata*, **133**, 195–218.
[2] Bhatia, Rajendra. 2007. *Positive definite matrices.* Princeton Series in Applied Mathematics. Princeton University Press, Princeton, NJ. [2015] paperback edition of the 2007 original [MR2284176].
[3] Cayley, A. 1841. On a theorem in the geometry of position. *Cambridge Math. Journal*, **2**, 267–271.
[4] Gromov, Misha. 2007. *Metric structures for Riemannian and non-Riemannian spaces.* English edn. Modern Birkhäuser Classics. Birkhäuser Boston, Inc. Based on the 1981 French original, with appendices by M. Katz, P. Pansu and S. Semmes. Translated from the French by Sean Michael Bates.
[5] Karliğa, Baki. 2004. Edge matrix of hyperbolic simplices. *Geom. Dedicata*, **109**, 1–6.
[6] Kowalick, R., Lafont, J.-F. and Minemyer, B. *Combinatorial systolic inequalities.* Preprint available at `http://arxiv.org/abs/1506.07121`.
[7] Lebedeva, Nina and Petrunin, Anton. 2010. Curvature bounded below: a definition a la Berg-Nikolaev. *Electron. Res. Announc. Math. Sci.*, **17**, 122–124.
[8] Menger, Karl. 1928. Untersuchungen über allgemeine Metrik. *Math. Ann.*, **100**(1), 75–163.
[9] Minemyer, B. 2017. Isometric embeddings of indefinite metric polyhedra. *Mosc. Math. J.*, **17**(1), 79–95.
[10] Minemyer, Barry. 2013. *Isometric embeddings of polyhedra.* Ph.D. thesis, State University of New York at Binghamton.
[11] Rivin, I. *Some observations on the simplex.* Preprint available on the `http://arxiv.org/abs/0308239v3`.

9
Hyperbolic Dimension and Decomposition Complexity

Andrew Nicas and David Rosenthal

Abstract

The aim of this chapter is to provide some new tools to aid the study of *decomposition complexity*, a notion introduced by Guentner, Tessera and Yu. In this chapter, three equivalent definitions for decomposition complexity are established. We prove that metric spaces with finite hyperbolic dimension have (weak) finite decomposition complexity, and we prove that the collection of metric families that are coarsely embeddable into Hilbert space is closed under decomposition. A method for showing that certain metric spaces do not have finite decomposition complexity is also discussed.

The *asymptotic dimension* of a metric space was introduced by Gromov [10] as a tool for studying the large scale geometry of groups. Interest in this concept intensified when Guoliang Yu proved the Novikov Conjecture for a finitely generated group G having finite asymptotic dimension as a metric space with a word-length metric and whose classifying space BG has the homotopy type of a finite complex, [17]. There are many geometrically interesting metric spaces that do not have finite asymptotic dimension. In order to study groups with infinite asymptotic dimension, Guentner, Tessera and Yu introduced the notion of *finite decomposition complexity*, abbreviated here to *FDC* [11]. Every countable group admits a proper left-invariant metric that is unique up to coarse equivalence. Guentner, Tessera and Yu showed that any countable subgroup of $GL(n, R)$, the group of invertible $n \times n$ matrices over an arbitrary commutative ring R has FDC [12]. Such a group can have infinite asymptotic dimension; for example, the wreath product $\mathbb{Z} \wr \mathbb{Z}$ (this finitely generated group can be realized as a subgroup of $GL(2, \mathbb{Z}[t, t^{-1}])$). The collection of countable groups with FDC contains groups with finite asymptotic dimension and has nice inheritance properties: it is closed under subgroups, extensions, free products with amalgamation, HNN extensions and countable direct unions. The FDC condition was introduced to study topological rigidity questions [11]. In this chapter we focus on finite decomposition complexity as a coarse geometric invariant.

The definition of finite decomposition complexity is somewhat tricky to work with, so it is advantageous to develop tools for determining whether or not a metric space has FDC. In this chapter we introduce some new tools for working with finite decomposition complexity (both the *strong* and *weak* forms), and try to give a feel for decomposition complexity by using these tools in several situations. Motivated by the equivalent definitions for finite asymptotic dimension, we provide analogous conditions that are equivalent to decomposition complexity, which we then use in the following two applications.

Buyalo and Schroeder introduced the *hyperbolic dimension* of a metric space (Definition 9.25) to study quasi-isometric embedding properties of negatively curved spaces. Cappadocia introduced the related notion of the *weak hyperbolic dimension* of a metric space (Definition 9.26). Hyperbolic dimension is an upper bound for weak hyperbolic dimension. We show in Corollary 9.30 that a metric space with weak hyperbolic dimension at most n is *n-decomposable* (Definition 9.3) over the collection of metric families with finite asymptotic dimension. This implies that such a metric space has weak FDC; if $n \leq 1$, then it has FDC.

In [6], Dadarlat and Guentner introduced the notion of a family of metric spaces that is *coarsely embeddable into Hilbert space*[1] (Definition 9.21). In Theorem 9.23, we show that if a metric family is n-decomposable over the collection of metric families that are coarsely embeddable into Hilbert space, then that metric family is also coarsely embeddable into Hilbert space. In other words, the collection of metric families that are coarsely embeddable into Hilbert space is stable under decomposition. This recovers the known fact that a metric space with (strong or weak) FDC is coarsely embeddable into Hilbert space.

Not all metric spaces satisfy the FDC condition. Clearly, any metric space that does not coarsely embed into Hilbert space will not have (strong or weak) FDC. Generalizing an example of Wu and Chen [16], we provide a tool that can be used to show that certain metric spaces

[1] Dadarlat and Guentner used the phrase "equi-uniformly embeddable" instead of "coarsely embeddable".

do not have (strong or weak) FDC. In Theorem 9.14 it is shown that if a metric space X admits a surjective *uniform expansion* and has weak finite decomposition complexity, then X has finite asymptotic dimension. Thus, as explained in Example 9.15, any infinite-dimensional normed linear space cannot have (strong or weak) FDC because such a space has infinite asymptotic dimension and admits a surjective uniform expansion.

In the final section of this chapter we recall some interesting open problems about decomposition complexity and suggest a few new ones.

9.1 Decomposition Complexity

Guentner, Tessera and Yu's concept of finite decomposition complexity was motivated by the following definition of finite asymptotic dimension.

Definition 9.1 Let n be a non-negative integer. The metric space (X, d) has *asymptotic dimension at most* n, $\operatorname{asdim} X \leq n$, if for every $r > 0$ there exists a cover $\mathcal{U}$ of X such that

(i) $\mathcal{U} = \mathcal{U}_0 \cup \mathcal{U}_1 \cup \cdots \cup \mathcal{U}_n$;
(ii) each $\mathcal{U}_i$, $0 \leq i \leq n$, is *r-disjoint*, i.e., $d(U, V) > r$ for every $U \neq V$ in $\mathcal{U}_i$; and
(iii) $\mathcal{U}$ is *uniformly bounded*, i.e., the *mesh* of $\mathcal{U}$,

$$\operatorname{mesh}(\mathcal{U}) = \sup\{\operatorname{diam}(U) \mid U \in \mathcal{U}\},$$

is finite.

If no such n exists, then $\operatorname{asdim} X = \infty$.

In the above definition, note that the cover $\mathcal{U}$ has *multiplicity* at most $n + 1$, i.e., every point of X is contained in at most $n + 1$ elements of $\mathcal{U}$. Also note that while $\mathcal{U}$ is not required to be an open cover, if $\operatorname{asdim} X < \infty$, then one can always choose $\mathcal{U}$ to be an open cover because of condition (ii).

The asymptotic dimension of a finitely generated group, G, is defined to be the asymptotic dimension of G considered as a metric space with the word-length metric associated with any finite set of generators. This is well-defined since asymptotic dimension is a coarse invariant and any two finite generating sets for G yield coarsely equivalent metric spaces. More generally, every countable group G admits a proper left-invariant metric that is unique up to coarse equivalence. Thus, asymptotic dimension can also be used as a coarse invariant for countable groups.

In order to generalize the definition of finite asymptotic dimension, it is useful to work with the notion of a *metric family*, a (countable) collection of metric spaces. A single metric space is viewed as a metric family with one element. A *subspace* of a metric family $\mathcal{X}$ is a metric family $\mathcal{Z}$ such that every element of $\mathcal{Z}$ is a metric subspace of some element of $\mathcal{X}$. For example, a cover $\mathcal{U}$ of a metric space X is a metric family, where each element of $\mathcal{U}$ is given the subspace metric inherited from X, and $\mathcal{U}$ is a subspace of the metric family $\{X\}$.

Definition 9.2 Let $r > 0$ and n be a non-negative integer. The metric family $\mathcal{X}$ is (r,n)*-decomposable* over the metric family $\mathcal{Y}$, denoted $\mathcal{X} \xrightarrow{(r,n)} \mathcal{Y}$, if for every X in $\mathcal{X}$, $X = X_0 \cup X_1 \cup \cdots \cup X_n$ such that for each i

$$X_i = \bigsqcup_{r\text{-disjoint}} X_{ij},$$

where each X_{ij} is in $\mathcal{Y}$.

Definition 9.3 Let n be a non-negative integer, and let $\mathfrak{C}$ be a collection of metric families. The metric family $\mathcal{X}$ is n*-decomposable* over $\mathfrak{C}$ if for every $r > 0$ $\mathcal{X}$ is (r,n)-decomposable over some metric family $\mathcal{Y}$ in $\mathfrak{C}$.

Following [11], we say that $\mathcal{X}$ is *weakly decomposable* over $\mathfrak{C}$ if $\mathcal{X}$ is n-decomposable over $\mathfrak{C}$ for some non-negative integer n, and $\mathcal{X}$ is *strongly decomposable* over $\mathfrak{C}$ if $\mathcal{X}$ is 1-decomposable over $\mathfrak{C}$.

Definition 9.4 A metric family $\mathcal{Z}$ is *bounded* if the diameters of the elements of $\mathcal{Z}$ are uniformly bounded, i.e., if $\sup\{\operatorname{diam}(Z) \mid Z \in \mathcal{Z}\} < \infty$. The collection of all bounded metric families is denoted by $\mathfrak{B}$.

Example 9.5 Let X be a metric space. The statement that the metric family $\{X\}$ is n-decomposable over $\mathfrak{B}$ is equivalent to the statement that $\operatorname{asdim}(X) \le n$.

The following definition is equivalent to Bell and Dranishnikov's definition of a collection of metric spaces having finite asymptotic dimension "uniformly" [1, Section 1].

Definition 9.6 Let n be a non-negative integer. The metric family $\mathcal{X}$ has *asymptotic dimension at most* n, denoted $\operatorname{asdim}(\mathcal{X}) \le n$, if $\mathcal{X}$ is n-decomposable over $\mathfrak{B}$.

Example 9.7 For each positive integer n, let $\mathcal{X}_n$ be the metric family of subsets of $\mathbb{R}^n$, with the Euclidean metric, consisting of open balls centered at the origin with positive integer radius. Then $\operatorname{asdim}(\mathcal{X}_n) = n$.

Proof Since $\mathcal{X}_n$ is a family of subspaces of $\mathbb{R}^n$, we have that $\operatorname{asdim}(\mathcal{X}_n) \leq \operatorname{asdim}(\mathbb{R}^n) = n$.

Suppose that $\operatorname{asdim}(\mathcal{X}_n) = \ell < n$. Then, for each integer $m \geq 1$, there exists a cover $\mathcal{U}_m$, which can be assumed to be an open cover of the open ball $B_m(0)$, with multiplicity at most $\ell + 1$, such that $\sup\{\operatorname{mesh}(\mathcal{U}_m) \mid m \geq 1\} = D < \infty$.

Let $\epsilon > 0$. Choose an integer k so that $k > D/\epsilon$. For $\lambda > 0$ and $A \subset \mathbb{R}^n$ let $\lambda A = \{\lambda a \mid a \in A\}$. Then $\mathcal{U} = \{\frac{1}{k} U \mid U \in \mathcal{U}_k\}$ is an open cover of $B_1(0)$ with $\operatorname{mesh}(\mathcal{U}) < \epsilon$ and multiplicity at most $\ell + 1$. Hence the covering dimension of $B_1(0)$ is at most ℓ, which contradicts the fact that the covering dimension of $B_1(0)$ is n. □

Definition 9.8 Let $\mathfrak{D}$ be the smallest collection of metric families containing $\mathfrak{B}$ that is closed under strong decomposition, and let $w\mathfrak{D}$ be the smallest collection of metric families containing $\mathfrak{B}$ that is closed under weak decomposition. A metric family in $\mathfrak{D}$ is said to have *finite decomposition complexity* (abbreviated to "FDC"), and a metric family in $w\mathfrak{D}$ is said to have *weak finite decomposition complexity* (abbreviated to "weak FDC").

Clearly, finite decomposition complexity implies weak finite decomposition complexity. The converse is unknown.

Question *[12, Question 2.2.6] Does weak finite decomposition complexity imply finite decomposition complexity?*

The base case of this question has an affirmative answer, namely, if $\operatorname{asdim} X < \infty$ then X has FDC [12, Theorem 4.1] (although even this case is difficult). Therefore, if $\mathfrak{A}$ is the collection of all metric families with finite asymptotic dimension, then we have the following sequence of inclusions of collections of metric families:

$$\mathfrak{A} \subset \mathfrak{D} \subset w\mathfrak{D}. \tag{9.1}$$

The collection of countable groups (considered as metric spaces with a proper left-invariant metric) in $\mathfrak{D}$ is quite large. It contains countable subgroups of $GL(n, R)$, where R is any commutative ring, countable subgroups of almost connected Lie groups, hyperbolic groups and elementary amenable groups. It is also closed under subgroups, extensions, free products with amalgamation, HNN extensions and countable direct unions [12].

The FDC and weak FDC conditions have important topological consequences. For example, a finitely generated group with weak FDC satisfies the *Novikov Conjecture*, and a metric space with (strong) FDC and bounded geometry satisfies the *Bounded Borel Conjecture* [11, 12]. These results were obtained by studying certain *assembly maps* in L-theory and topological K-theory. The assembly map in algebraic K-theory has been studied for groups with FDC by several authors, including Ramras, Tessera and Yu [15], Kasprowski [13] and Goldfarb [8].

There is an equivalent description of FDC, and weak FDC, in terms of a *metric decomposition game* [12, Theorem 2.2.3], that is useful for understanding the proofs of many of the inheritance properties mentioned above. The metric decomposition game has two players, a defender and a challenger. The game begins with a metric family $\mathcal{X} = \mathcal{Y}_0$. On the first turn, the challenger declares a positive integer r_1 and the defender must produce a (r_1, n_1)-decomposition of $\mathcal{Y}_0$ over a new metric family $\mathcal{Y}_1$. On the second turn, the challenger declares a positive integer r_2 and the defender must produce an (r_2, n_2)-decomposition of $\mathcal{Y}_1$ over a new metric family $\mathcal{Y}_2$. The game continues in this manner, ending if and when the defender produces a bounded family. In this case the defender has won. A winning strategy is a set of instructions that, if followed by the defender, will guarantee a win for any possible requests made by the challenger. The family $\mathcal{X}$ has weak FDC if a winning strategy exists and strong FDC if, additionally, the strategy always allows for $n_j = 1$ in the defender's response.

Next, we recall some terminology introduced in [12] that generalizes basic notions from the coarse geometry of metric spaces to metric families.

Let $\mathcal{X}$ and $\mathcal{Y}$ be metric families. A *map of families*, $F : \mathcal{X} \to \mathcal{Y}$, is a collection of functions $F = \{f : X \to Y\}$, where $X \in \mathcal{X}$ and $Y \in \mathcal{Y}$, such that every $X \in \mathcal{X}$ is the domain of at least one f in F. The *inverse image of* $\mathcal{Z}$ *under* F is the subspace of $\mathcal{X}$ given by $F^{-1}(\mathcal{Z}) = \{f^{-1}(Z) \mid Z \in \mathcal{Z}, f \in F\}$.

Definition 9.9 A map of metric families, $F : \mathcal{X} \to \mathcal{Y}$, is a *coarse embedding* if there exist non-decreasing functions $\delta, \rho : [0, \infty) \to [0, \infty)$, with $\lim_{t\to\infty} \delta(t) = \infty = \lim_{t\to\infty} \rho(t)$, such that for every $f : X \to Y$ in

F and every $x, y \in X$,

$$\delta\big(d_X(x,y)\big) \le d_Y\big(f(x), f(y)\big) \le \rho\big(d_X(x,y)\big).$$

One can think of a coarse embedding of metric families as a collection of "uniform" coarse embeddings, in the sense that they have a common δ and ρ. The easiest example of a coarse embedding of metric families is the inclusion of a subspace $\mathcal{Z}$ of $\mathcal{Y}$ into $\mathcal{Y}$.

Definition 9.10 A map of metric families, $F : \mathcal{X} \to \mathcal{Y}$, is a *coarse equivalence* if for each $f : X \to Y$ in F there is a map $g_f : Y \to X$ such that:

(i) the collection $G = \{g_f\}$ is a coarse embedding from $\mathcal{Y}$ to $\mathcal{X}$; and
(ii) the composites $f \circ g_f$ and $g_f \circ f$ are *uniformly close* to the identity maps id_Y and id_X, respectively, in the sense that there is a constant $C > 0$ with

$$d_Y\big(y, f \circ g_f(y)\big) \le C \text{ and } d_X\big(x, g_f \circ f(x)\big) \le C,$$

for every $f : X \to Y$ in F, $x \in X$, and $y \in Y$.

Definition 9.11 A collection of metric families, $\mathfrak{C}$, is *closed under coarse embeddings* if every metric family $\mathcal{X}$ that coarsely embeds into a metric family $\mathcal{Y}$ in $\mathfrak{C}$ is also a metric family in $\mathfrak{C}$.

Guentner, Tessera and Yu proved that $\mathfrak{D}$ and $w\mathfrak{D}$ are each closed under coarse embeddings [12, Coarse Invariance 3.1.3]. It is straightforward to check that the following collections of metric families are also closed under coarse embeddings.

Example 9.12 Collections of metric families that are closed under coarse embeddings.

- $\mathfrak{B}$, the collection of bounded metric families.
- $\mathfrak{A}$, the collection of metric families with finite asymptotic dimension.
- $\mathfrak{A}_n$, the collection of metric families with asymptotic dimension at most n.
- $\mathfrak{H}$, the collection of metric families that are *coarsely embeddable into Hilbert space* (see Definition 9.21).

The following is similar to [12, Coarse Invariance 3.1.3].

Theorem 9.13 *Let $\mathcal{X}$ and $\mathcal{Y}$ be metric families, and let $\mathfrak{C}$ be a collection of metric families that is closed under coarse embeddings. If $\mathcal{X}$*

coarsely embeds into $\mathcal{Y}$ and $\mathcal{Y}$ is n-decomposable over $\mathfrak{C}$, then $\mathcal{X}$ is n-decomposable over $\mathfrak{C}$. In particular, if $\mathcal{X}$ is coarsely equivalent to $\mathcal{Y}$, then $\mathcal{X}$ is n-decomposable over $\mathfrak{C}$ if and only if $\mathcal{Y}$ is n-decomposable over $\mathfrak{C}$.

Proof Let $F : \mathcal{X} \to \mathcal{Y}$ be a coarse embedding, and let $r > 0$ be given. We must find a metric family $\mathcal{X}'$ in $\mathfrak{C}$ such that $\mathcal{X}$ is (r, n)-decomposable over $\mathcal{X}'$. Since $\mathcal{Y}$ is n-decomposable over $\mathfrak{C}$, there is a metric family $\mathcal{Y}'$ such that $\mathcal{Y}$ is $(\rho(r), n)$-decomposable over $\mathcal{X}'$, where ρ is as in Definition 9.9. It is straightforward to show that $\mathcal{X}$ is (r, n)-decomposable over $\mathcal{X}' = F^{-1}(\mathcal{Y}')$. Note that F restricts to a coarse embedding from $F^{-1}(\mathcal{Y}')$ to $\mathcal{Y}'$. Since $\mathfrak{C}$ is closed under coarse embeddings, we are done. □

The following observation about decomposition is useful.

Remark If $\mathcal{X}$, $\mathcal{Y}$ and $\mathcal{Z}$ are metric families and $\mathcal{X} \xrightarrow{(r,m)} \mathcal{Y} \xrightarrow{(s,n)} \mathcal{Z}$, then $\mathcal{X} \xrightarrow{(t,p)} \mathcal{Z}$ where $t = \min(r, s)$ and $p = (m+1)(n+1) - 1$. In particular, this shows that if $\mathcal{X}$ is m-decomposable over $\mathfrak{A}_n$, then $\mathcal{X}$ has asymptotic dimension at most $(m+1)(n+1) - 1$.

Let (X, d_X) be a metric space and $\lambda > 1$. A *uniform expansion of X with expansion factor* λ is a map $T\colon X \to X$ such that $d_X(T(x), T(y)) = \lambda\, d_X(x, y)$ for all $x, y \in X$. The following proposition, generalizing [16, Example 2.2], can be used to show that certain spaces do *not* have weak finite decomposition complexity.

Theorem 9.14 *Let (X, d_X) be a metric space that admits a surjective uniform expansion. If X has weak finite decomposition complexity then X has finite asymptotic dimension.*

Proof Assume that the metric space (X, d_X) has weak FDC and that $T\colon X \to X$ is a surjective uniform expansion of X with expansion factor $\lambda > 1$. By the analog of [11, Theorem 2.4] for weak FDC (while [11, Theorem 2.4] is stated for FDC, the proof there readily adapts to weak FDC), there exists a finite sequence (r_i, n_i), $i = 1, \ldots, m$, where each $r_i > 0$ and the n_is are positive integers, together with metric families $\mathcal{Y}_i$, $i = 1, \ldots, m$, such that

$$X \xrightarrow{(r_1,n_1)} \mathcal{Y}_1 \xrightarrow{(r_2,n_2)} \mathcal{Y}_2 \longrightarrow \cdots \xrightarrow{(r_m,n_m)} \mathcal{Y}_m$$

and $\mathcal{Y}_m \in \mathfrak{B}$. (This is one winning round of the "decomposition game".) By Remark 9.1, we have $X \xrightarrow{(r,n)} \mathcal{Y}_m$, where $r = \min\{r_1, \ldots, r_m\}$ and $n = (n_1+1)(n_2+1)\cdots(n_m+1) - 1$. For any positive integer k, let

$T^k(\mathcal{Y}_m) = \{T^k(Y) \mid Y \in \mathcal{Y}_m\}$. Notice that $T^k(\mathcal{Y}_m) \in \mathfrak{B}$. Since T is surjective, $T^k(X) = X$. Hence,

$$\{X\} = \{T^k(X)\} \xrightarrow{(\lambda^k r, n)} T^k(\mathcal{Y}_m).$$

Since $\lambda > 1$, we have $\lambda^k r \to \infty$ as $k \to \infty$. It follows that $\{X\}$ is n-decomposable over $\mathfrak{B}$. That is, X has finite asymptotic dimension. □

Example 9.15 Let $(V, \|\cdot\|)$ be any infinite-dimensional normed linear space. Then $T(x) = 2x$ is a uniform expansion of V, with expansion factor 2, where the metric is $d(x, y) = \|x - y\|$. Clearly, T is surjective. Note that any real n-dimensional vector subspace of V has asymptotic dimension n and so V has infinite asymptotic dimension. It follows from Theorem 9.14 that (V, d) cannot have weak FDC.

Example 9.16 The condition in Theorem 9.14 that the uniform expansion T is surjective cannot be omitted. Consider $X = \bigoplus_{i=1}^{\infty} \mathbb{Z}$ with the proper metric $d_X\big((x_i), (y_i)\big) = \sum_{i=1}^{\infty} i \cdot |x_i - y_i|$. Then there is a uniform expansion $T\big((x_i)\big) = (2x_i)$ of (X, d_X), with expansion factor 2, but T is not surjective. Although (X, d_X) has infinite asymptotic dimension, it has FDC (see [11, Example 2.5]) and hence weak FDC.

Now consider $Y = \bigoplus_{i=1}^{\infty} \mathbb{R}$ equipped with the metric $d_Y\big((x_i), (y_i)\big) = \sum_{i=1}^{\infty} i \cdot |x_i - y_i|$. Then X is a metric subspace of Y, and for each $n \in \mathbb{N}$, the subspace $\bigoplus_{i=1}^{n} \mathbb{Z}$ of X is coarsely equivalent to the subspace $\bigoplus_{i=1}^{n} \mathbb{R}$ of Y. Nevertheless, X is *not* coarsely equivalent to Y, since X does not have weak FDC by Example 9.15.

If $\mathcal{X}$ is a metric family and $N = \sup\{\text{asdim}(X) \mid X \in \mathcal{X}\}$, then clearly $N \leq \text{asdim}(\mathcal{X})$. Equality often does *not* hold. For example, consider the space $Z = \bigoplus_{i=1}^{\infty} \mathbb{R}$ with the Euclidean metric and the metric family $\mathcal{X} = \{B_r(0) \mid r = 1, 2, \ldots\}$ of open balls in Z. For each positive integer n, let $\mathcal{X}_n = \{B_r(0) \cap \mathbb{R}^n \mid r = 1, 2, \ldots\}$, where $\mathbb{R}^n$ denotes the metric subspace $\bigoplus_{i=1}^{n} \mathbb{R}$ in Z. Then, by Example 9.7, $\text{asdim}(\mathcal{X}_n) = n$. Therefore, $n = \text{asdim}(\mathcal{X}_n) \leq \text{asdim}(\mathcal{X})$ for every positive integer n, and so $\text{asdim}(\mathcal{X}) = \infty$, whereas $\text{asdim}(B_r(0)) = 0$ for each r.

However, as was pointed out to us by Daniel Kasprowski, for every countable discrete group G equipped with a proper left-invariant metric, the family of finite subgroups of G does have asymptotic dimension zero as a metric family. The following proposition is a generalization of this fact.

Proposition 9.17 *Let G be a countable discrete group equipped with* *a proper left-invariant metric d. Let $\mathcal{F}$ be a non-empty collection of subgroups of G that is closed under taking subgroups. If* $\operatorname{asdim}(H) \leq k$ *for every H in $\mathcal{F}$, where H is considered as a metric subspace of G, then* $\operatorname{asdim}(\mathcal{F}) \leq k$.

Proof Let $r > 0$ be given. For each H in $\mathcal{F}$, let S_H be the subgroup of H generated by $H \cap B_r(e)$, where e is the identity element of G. Let $\mathcal{U}_H$ be the set of left cosets of S_H in H. Then, for every $x, y \in H$,

$$d(x,y) \leq r \;\Leftrightarrow\; x^{-1}y \in B_r(e) \;\Rightarrow\; x^{-1}y \in S_H.$$

Thus, $\mathcal{U}_H$ is an r-disjoint, 0-dimensional cover of H. Let $\mathcal{Y}$ be the metric family $\bigcup_{H \in \mathcal{F}} \mathcal{U}_H$. Then, $\mathcal{F}$ is $(r, 0)$-decomposable over $\mathcal{Y}$. Since $\mathcal{U}_H$ is coarsely equivalent to $\{S_H\}$, it follows that $\mathcal{Y}$ is coarsely equivalent to $\{S_H \mid H \in \mathcal{F}\}$, which is a finite set since d is a proper metric. Therefore, $\operatorname{asdim}(\mathcal{Y}) = \operatorname{asdim}\left(\{S_H \mid H \in \mathcal{F}\}\right) \leq k$. Thus, $\mathcal{F}$ is 0-decomposable over $\mathfrak{A}_k$, the collection of all metric families that have asymptotic dimension at most k. It follows from Remark 9.1 that $\operatorname{asdim}(\mathcal{F}) \leq k$. □

9.2 Equivalent Definitions of Decomposability

In this section we provide three alternative definitions for a metric family $\mathcal{X}$ to be n-decomposable over a collection of metric families $\mathfrak{C}$. We show that they are all equivalent to Definition 9.3, provided $\mathfrak{C}$ is closed under coarse embeddings. When $\mathfrak{C} = \mathfrak{B}$ (the collection of all bounded metric families) and $\mathcal{X}$ consists of a single metric space, each of our definitions reduces to one of the standard definitions for finite asymptotic dimension.

Recall that the multiplicity of a covering $\mathcal{U}$ of a metric space X is the largest integer m such that every point of X is contained in at most m elements of $\mathcal{U}$. Given $d > 0$, the *d-multiplicity* of $\mathcal{U}$ is the largest integer m such that every open d-ball, $B_d(x)$, in X is contained in at most m elements of $\mathcal{U}$. The *Lebesgue number* of $\mathcal{U}$, $L(\mathcal{U})$, is at least $\lambda > 0$ if every $B_\lambda(x)$ in X is contained in some element of $\mathcal{U}$. A *uniform simplicial complex* K is a simplicial complex equipped with the ℓ^1-metric. That is, every element $x \in K$ can be uniquely written as $x = \sum_{v \in K^{(0)}} x_v \cdot v$, where $K^{(0)}$ is the vertex set of K, each $x_v \in [0,1]$, $x_v = 0$ for all but finitely many $v \in K^{(0)}$, and $\sum_{v \in K^{(0)}} x_v = 1$. Then the ℓ^1-metric is defined by $d^1(x,y) = \sum_{v \in K^{(0)}} |x_v - y_v|$. The *open star* of a vertex $v \in K^{(0)}$ is the set $\operatorname{star}(v) = \{x \in K \mid x_v \neq 0\}$. If there exists an integer m such that for every $x \in K$ the set $\{v \in K^{(0)} \mid x_v \neq 0\}$ has cardinality at most m, then the *dimension of* K, $\dim(K)$, is at most m. If no such m exists, then $\dim(K) = \infty$.

In what follows, let $\mathcal{X} = \{X_\alpha \mid \alpha \in I\}$ be a metric family, where I is a countable indexing set, and let $\mathfrak{C}$ be a collection of metric families. Let n be a non-negative integer.

Condition (A) For every $d > 0$, there exists a cover $\mathcal{V}_\alpha$ of X_α, for each $\alpha \in I$, such that:

(i) the d-multiplicity of $\mathcal{V}_\alpha$ is at most $n+1$ for every $\alpha \in I$; and
(ii) $\bigcup_{\alpha \in I} \mathcal{V}_\alpha$ is a metric family in $\mathfrak{C}$.

Condition (B) For every $\lambda > 0$, there exists a cover $\mathcal{U}_\alpha$ of X_α, for each $\alpha \in I$, such that:

(i) the multiplicity of $\mathcal{U}_\alpha$ is at most $n+1$ for every $\alpha \in I$;
(ii) the Lebesgue number $L(\mathcal{U}_\alpha) \geq \lambda$ for every $\alpha \in I$; and
(iii) $\bigcup_{\alpha \in I} \mathcal{U}_\alpha$ is a metric family in $\mathfrak{C}$.

Condition (C) For every $\varepsilon > 0$, there exists a uniform simplicial complex K_α and an ε-Lipschitz map $\varphi_\alpha : X_\alpha \to K_\alpha$, for each $\alpha \in I$, such that:

(i) $\dim(K_\alpha) \leq n$ for every $\alpha \in I$; and
(ii) $\bigcup_{\alpha \in I} \{\varphi_\alpha^{-1}(\mathrm{star}(v)) \mid v \in K_\alpha^{(0)}\}$ is a metric family in $\mathfrak{C}$.

Proposition 9.18 *Let $\mathcal{X}$ be a metric family and $\mathfrak{C}$ be a collection of metric families that is closed under coarse embeddings. Then Conditions (A) and (B) are each equivalent to Definition 9.3.*

Proof For notational convenience we prove the proposition when $\mathcal{X}$ consists of a single metric space X. The proof for a general metric family is a straightforward generalization of this case.

Suppose that X is n-decomposable over $\mathfrak{C}$. Let $d > 0$ be given. Then, there is a metric family $\mathcal{Y}$ in $\mathfrak{C}$ and a decomposition $X = X_0 \cup X_1 \cup \cdots \cup X_n$ such that, for each i

$$X_i = \bigsqcup_{2d\text{-disjoint}} X_{ij},$$

where each X_{ij} is in $\mathcal{Y}$. Thus, the cover $\mathcal{V} = \{X_{ij}\}$ of X is a subspace of $\mathcal{Y}$ and has d-multiplicity less than or equal to $n+1$. Since $\mathfrak{C}$ is closed under coarse embeddings, $\mathcal{V}$ is also in $\mathfrak{C}$ and Condition (A) is satisfied.

Suppose that Condition (A) is satisfied for n with respect to $\mathfrak{C}$ and let

$\lambda > 0$ be given. There exists a cover $\mathcal{V}$ of X that is a metric family in $\mathfrak{C}$ and has λ-multiplicity less than or equal to $n+1$. Let $\mathcal{U} = \{V^\lambda \mid V \in \mathcal{V}\}$, where V^λ is the set of points in X whose distance from V is at most λ. Then the Lebesgue number $L(\mathcal{U}) \geq \lambda$. Given $x \in X$, the ball of radius λ around x intersects at most $n+1$ elements of $\mathcal{V}$, since the λ-multiplicity of $\mathcal{V}$ is at most $n+1$. This implies that at most $n+1$ elements of $\mathcal{U}$ contain x, i.e., the multiplicity of $\mathcal{U}$ is at most $n+1$. Since $\mathcal{U}$ is coarsely equivalent to $\mathcal{V}$ and $\mathfrak{C}$ is closed under coarse embeddings (and hence under coarse equivalences), Condition (B) is satisfied.

Suppose that Condition (B) is satisfied for n with respect to $\mathfrak{C}$ and let $r > 0$ be given. We follow an argument analogous to the one in [9, Theorem 9] to show that Condition (B) implies Definition 9.3. There exists a cover $\mathcal{U}$ of X such that $\mathcal{U}$ has multiplicity at most $n+1$, $L(\mathcal{U}) \geq (n+1)r$, and $\mathcal{U}$ is in $\mathfrak{C}$. Given $d > 0$ and $U \subset X$, let $\mathrm{Int}_d(U) = \{x \in X \mid B_d(x) \subset U\}$. Note that if $d_1 \leq d_2$, then $\mathrm{Int}_{d_2}(U) \subseteq \mathrm{Int}_{d_1}(U)$. Also note that if $a \in \mathrm{Int}_d(U) \cap \mathrm{Int}_d(V)$, then $a \in \mathrm{Int}_d(U \cap V)$. Now, for each $i \in \{0, \ldots, n\}$, define

$$\begin{aligned}
\mathcal{U}_i &= \{U_0 \cap U_1 \cap \cdots \cap U_i \mid U_0, U_1, \ldots, U_i \in \mathcal{U} \text{ are distinct}\}, \\
S_i &= \textstyle\bigcup_{U \in \mathcal{U}_i} \mathrm{Int}_{(n+1-i)r}(U), \\
X_i &= \textstyle\bigsqcup_{U \in \mathcal{U}_i} \mathrm{Int}_{(n+1-i)r}(U) \setminus S_{i+1}.
\end{aligned}$$

Since $\mathcal{U}$ has multiplicity at most $n+1$ and has Lebesgue number $L(\mathcal{U}) \geq (n+1)r$, it follows that $X = X_0 \cup X_1 \cup \cdots \cup X_n$. Furthermore, since each $\mathrm{Int}_{(n+1-i)r}(U) \setminus S_{i+1}$ is contained in some element of $\mathcal{U}$ and $\mathfrak{C}$ is closed under coarse embeddings, the metric family

$$\{\mathrm{Int}_{(n+1-i)r}(U) \setminus S_{i+1} \mid 0 \leq i \leq n \text{ and } U \in \mathcal{U}_i\}$$

is in $\mathfrak{C}$. It remains to show that in fact each X_i is an r-disjoint union. We do this by contradiction. Given i, suppose that $\mathrm{Int}_{(n+1-i)r}(U) \setminus S_{i+1} \neq \mathrm{Int}_{(n+1-i)r}(V) \setminus S_{i+1}$, where $U = U_0 \cap U_1 \cap \cdots \cap U_i$ and $V = V_0 \cap V_1 \cap \cdots \cap V_i$, and that there exist $a \in \mathrm{Int}_{(n+1-i)r}(U) \setminus S_{i+1}$ and $b \in \mathrm{Int}_{(n+1-i)r}(V) \setminus S_{i+1}$ with $d(a,b) \leq r$. Then $a \in \left(\mathrm{Int}_{(n+1-i)r}(V)\right)^r$ and $b \in \left(\mathrm{Int}_{(n+1-i)r}(U)\right)^r$. Notice that for each natural number k, the r-neighborhood

$$\begin{aligned}
&\left(\mathrm{Int}_{(k+1)r}(U)\right)^r \\
&\qquad = \{y \in X \mid \exists\, x \text{ such that } B_{(k+1)r}(x) \subset U \text{ and } d(y,x) \leq r\}
\end{aligned}$$

is contained in $\text{Int}_{kr}(U) = \{y \in X \mid B_{kr}(y) \subset U\}$. Therefore, $a \in \text{Int}_{(n-i)r}(V)$ and $b \in \text{Int}_{(n-i)r}(U)$. Thus,

$$a, b \in \text{Int}_{(n-i)r}(U) \cap \text{Int}_{(n-i)r}(V) \subset \text{Int}_{(n-i)r}(U \cap V).$$

Since $\text{Int}_{(n+1-i)r}(U) \setminus S_{i+1} \neq \text{Int}_{(n+1-i)r}(V) \setminus S_{i+1}$, we see that the set $\{U_0, \ldots, U_i, V_0, \ldots, V_i\}$ has at least $i+2$ elements, which implies that a and b are both in

$$\text{Int}_{(n-i)r}(U \cap V) = \text{Int}_{(n-i)r}(U_0 \cap \cdots \cap U_i \cap V_0 \cap \cdots \cap V_i) \subset S_{i+1}.$$

But this contradicts the assumption that $a \in \text{Int}_{(n+1-i)r}(U) \setminus S_{i+1}$ and $b \in \text{Int}_{(n+1-i)r}(V) \setminus S_{i+1}$. Therefore, X is n-decomposable over $\mathfrak{C}$. □

Lemma 9.19 *Let n be a non-negative integer. Then for every uniform simplicial complex K with $\dim(K) \leq n$, the cover $\mathcal{V}_K$ of K consisting of the open stars of vertices in K has Lebesgue number $L(\mathcal{V}_K) \geq \frac{1}{n+1}$.*

Proof Let K be a uniform simplicial complex of dimension n, and let $x \in K$ be given. Then $x = \sum_{v \in K^{(0)}} x_v \cdot v$, where there are at most $n+1$ vertices v with $x_v \neq 0$ and $\sum_{v \in K^{(0)}} x_v = 1$. There is a $v \in K^{(0)}$ with $x_v \geq \frac{1}{n+1}$. If $y \in K$ is not in the open star of v, $\text{star}(v)$, then $y_v = 0$ and $d^1(x, y) \geq \frac{1}{n+1}$. Therefore, the open ball of radius $\frac{1}{n+1}$ centered at x is completely contained in $\text{star}(v)$. Thus, the cover $\mathcal{V}_K$ of K consisting of the open stars of vertices in K has Lebesgue number $L(\mathcal{V}_K) \geq \frac{1}{n+1}$. □

Proposition 9.20 *Let $\mathcal{X}$ be a metric family and $\mathfrak{C}$ be a collection of metric families that is closed under coarse embeddings. Then Condition (C) is equivalent to Definition 9.3.*

Proof For notational convenience we prove the proposition when $\mathcal{X}$ consists of a single metric space X. The proof for a general metric family is a straightforward generalization of this case.

By Proposition 9.18, it suffices to prove that Condition (C) is equivalent to Condition (B). We follow [1, Assertion 2] to show that Condition (C) implies Condition (B), and we follow [2, Theorem 1] to show that Condition (B) implies Condition (C).

Assume that X satisfies Condition (C) for n with respect to $\mathfrak{C}$. Let $r > 0$ be given. Then, by Lemma 9.19, there is a uniform simplicial complex K of dimension n and a $\frac{1}{(n+1)r}$-Lipschitz map $\varphi : X \to K$. Since $\dim(K) = n$, the cover $\mathcal{U} = \{\varphi^{-1}(\text{star}(v)) \mid v \in K^{(0)}\}$ of X has multiplicity at most $n+1$ and Lebesgue number $L(\mathcal{U}) > r$. By assumption, the metric family $\{\varphi^{-1}(\text{star}(v)) \mid v \in K^{(0)}\}$ is in $\mathfrak{C}$. Thus, X satisfies Condition (B).

Now assume that X satisfies Condition (B) for n with respect to $\mathfrak{C}$. Let $\varepsilon > 0$ be given. Then there is a cover $\mathcal{U}$ of X that is a metric family in $\mathfrak{C}$, has multiplicity at most $n+1$ and has Lebesgue number $L(\mathcal{U}) \geq \lambda = \frac{(2n+2)(2n+3)}{\varepsilon}$. Note that, because $L(\mathcal{U}) > 0$ and $\mathfrak{C}$ is closed under coarse embeddings, we may additionally assume, without loss of generality, that $\mathcal{U}$ is an open covering of X. For each $U \in \mathcal{U}$, define $\varphi_U : X \to [0,1]$ by

$$\varphi_U(x) = \frac{d(x, U^c)}{\sum_{V \in \mathcal{U}} d(x, V^c)},$$

where U^c is the complement of U in X. Let $K = \text{Nerve}(\mathcal{U})$ equipped with the uniform metric. Since the multiplicity of $\mathcal{U}$ is at most $n+1$, $\dim(K) \leq n$. Define the map $\varphi : X \to K$ by

$$\varphi(x) = \sum_{U \in \mathcal{U}} \varphi_U(x) \cdot [U],$$

where $[U]$ denotes the vertex of K defined by U. Note that given a vertex $[V]$ in K, $\varphi^{-1}(\text{star}([V])) \subset V$, since $\varphi(x)$ is in the open star of $[V]$ if and only if $\varphi_V(x) \neq 0$, and this implies that x is in V. Therefore, the metric family $\{\varphi^{-1}(\text{star}([V])) \mid [V] \in K^{(0)}\} \subset \mathcal{U}$ is in $\mathfrak{C}$ since $\mathfrak{C}$ is closed under coarse embeddings.

It remains to show that φ is ε-Lipschitz. Since $L(\mathcal{U}) \geq \lambda$, it follows that $\sum_{V \in \mathcal{U}} d(x, V^c) \geq \lambda$. Also note that for every $x, y \in X$ and $U \in \mathcal{U}$, the triangle inequality implies

$$\left| d(x, U^c) - d(y, U^c) \right| \leq d(x, y).$$

Thus,

$$\begin{aligned}
\left| \varphi_U(x) - \varphi_U(y) \right| &= \left| \frac{d(x, U^c)}{\sum_{V \in \mathcal{U}} d(x, V^c)} - \frac{d(y, U^c)}{\sum_{V \in \mathcal{U}} d(y, V^c)} \right| \\
&\leq \frac{\left| d(x, U^c) - d(y, U^c) \right|}{\sum_{V \in \mathcal{U}} d(x, V^c)} + \left| \frac{d(y, U^c)}{\sum_{V \in \mathcal{U}} d(x, V^c)} - \frac{d(y, U^c)}{\sum_{V \in \mathcal{U}} d(y, V^c)} \right|
\end{aligned}$$

which is less than or equal to

$$\frac{d(x, y)}{\sum_{V \in \mathcal{U}} d(x, V^c)} + \frac{d(y, U^c) \cdot \sum_{V \in \mathcal{U}} \left| d(x, V^c) - d(y, V^c) \right|}{\left(\sum_{V \in \mathcal{U}} d(x, V^c) \right) \left(\sum_{V \in \mathcal{U}} d(y, V^c) \right)},$$

which is less than or equal to

$$\frac{1}{\lambda}\, d(x,y) + \frac{1}{\lambda}\Big(\sum_{V \in \mathcal{U}} \big| d(x,V^c) - d(y,V^c) \big| \Big)$$
$$\le \frac{1}{\lambda}\, d(x,y) + \frac{1}{\lambda}\, 2(n+1)\, d(x,y)$$
$$= \frac{1}{\lambda}\,(2n+3)\, d(x,y).$$

Therefore,

$$\begin{aligned} d^1(\varphi(x), \varphi(y)) &= \sum_{U \in \mathcal{U}} \big| \varphi_U(x) - \varphi_U(y) \big| \\ &\le 2(n+1) \left(\frac{1}{\lambda}\,(2n+3)\, d(x,y) \right) = \varepsilon\, d(x,y). \end{aligned}$$

This completes the proof. □

The equivalent definitions for decomposability give us more tools to work with. For instance, consider the collection of metric families that are *coarsely embeddable into Hilbert space*, defined below. The notion of a metric family that is coarsely embeddable into Hilbert space was introduced by Dadarlat and Guentner in [6], although they called it a "family of metric spaces that is equi-uniformly embeddable".

Definition 9.21 A metric family $\mathcal{X} = \{X_\alpha \mid \alpha \in I\}$ is *coarsely embeddable into Hilbert space* if there is a family of Hilbert spaces $\mathcal{H} = \{H_\alpha \mid \alpha \in I\}$ and a map of metric families $F = \{F_\alpha : X_\alpha \to H_\alpha \mid \alpha \in I\}$ such that $F : \mathcal{X} \to \mathcal{H}$ is a coarse embedding. The collection of all metric families that are coarsely embeddable into Hilbert space is denoted by $\mathfrak{H}$.

In [7], Dadarlat and Guentner proved the following.

Proposition 9.22 *[7, Proposition 2.3] A metric family $\mathcal{X} = \{X_\alpha \mid \alpha \in I\}$ is in $\mathfrak{H}$ if and only if for every $R > 0$ and $\varepsilon > 0$ there exists a family of Hilbert spaces $\mathcal{H} = \{H_\alpha \mid \alpha \in I\}$ and a map of metric families $\xi = \{\xi_\alpha : X_\alpha \to H_\alpha \mid \alpha \in I\}$ such that*

(i) $\|\xi_\alpha(x)\| = 1$, *for all* $x \in X_\alpha$ *and* $\alpha \in I$;
(ii) $\forall \alpha \in I$, $\forall x, x' \in X_\alpha$, $d_\alpha(x,x') \le R \Rightarrow \|\xi_\alpha(x) - \xi_\alpha(x')\| \le \varepsilon$;
(iii) $\lim_{S \to \infty} \sup_{\alpha \in I} \sup \big\{ \big| \langle \xi_\alpha(x), \xi_\alpha(x') \rangle \big| : d_\alpha(x,x') \ge S,\ x, x' \in X_\alpha \big\} = 0.$ □

Proposition 9.20 enables us to make use of Dadarlat and Guentner's work to prove the following theorem.

Theorem 9.23 *The collection $\mathfrak{H}$ of metric families that are coarsely embeddable into Hilbert space is stable under weak decomposition. That is, if a metric family $\mathcal{X}$ is n-decomposable over $\mathfrak{H}$, then $\mathcal{X}$ is in $\mathfrak{H}$.*

Proof In the light of Proposition 9.20, all of the ingredients for the proof of this theorem are contained in [7]. The argument is organized as follows.

Let $\mathcal{X} = \{X_\alpha \mid \alpha \in I\}$ be a metric family that is n-decomposable over $\mathfrak{H}$. We will use Proposition 9.22 to prove that $\mathcal{X}$ is in $\mathfrak{H}$. Let $R > 0$ and $\varepsilon > 0$ be given. Since $\mathcal{X}$ is n-decomposable over $\mathfrak{H}$, the proof of Proposition 9.20 implies that for each $\alpha \in I$ there is a cover $\mathcal{U}_\alpha = \{U_{\alpha,j}\}_{j \in J_\alpha}$ of X_α, for some indexing set J_α, and a family of maps $\varphi_\alpha = \{\varphi_{\alpha,j} : X_\alpha \to [0,1] \mid j \in J_\alpha\}$ such that

(a) $\sum_{j \in J_\alpha} \varphi_{\alpha,j}(x) = 1$, for all $x \in X_\alpha$;
(b) $\varphi_{\alpha,j}(x) = 0$ if $x \notin U_{\alpha,j}$;
(c) $\forall x, y \in X_\alpha$, $d_\alpha(x,y) \leq R \Rightarrow \sum_{j \in J} \left|\varphi_{\alpha,j}(x) - \varphi_{\alpha,j}(y)\right| \leq \frac{\varepsilon^2}{4}$;
(d) the metric family $\{U_{\alpha,j} \mid \alpha \in I, j \in J_\alpha\}$ is in $\mathfrak{H}$.

The metric family $\{U^R_{\alpha,j} \mid \alpha \in I, j \in J_\alpha\}$, where $U^R_{\alpha,j} = \{x \in X_\alpha \mid d_\alpha(x, U_{\alpha,j}) \leq R\}$, is coarsely equivalent to the metric family $\{U_{\alpha,j} \mid \alpha \in I, j \in J_\alpha\}$. Therefore, since $\{U_{\alpha,j} \mid \alpha \in I, j \in J_\alpha\}$ is in $\mathfrak{H}$, so is $\{U^R_{\alpha,j} \mid \alpha \in I, j \in J_\alpha\}$. By Proposition 9.22, there exists a family of Hilbert spaces $\mathcal{H} = \{H_{\alpha,j} \mid \alpha \in I, j \in J_\alpha\}$ and a map of metric families $\xi = \{\xi_{\alpha,j} : U^R_{\alpha,j} \to H_{\alpha,j} \mid \alpha \in I, j \in J_\alpha\}$ satisfying

(i) $\|\xi_{\alpha,j}(x)\| = 1$, for all $x \in U^R_{\alpha,j}$;
(ii) $\sup\left\{\|\xi_{\alpha,j}(x) - \xi_{\alpha,j}(y)\| : d_\alpha(x,y) \leq R, x, y \in U^R_{\alpha,j}\right\} \leq \varepsilon/2$, for all $\alpha \in I$, $j \in J_\alpha$;
(iii) $\lim\limits_{S \to \infty} \sup\limits_{\alpha \in I, j \in J_\alpha} \sup\left\{\left|\langle \xi_{\alpha,j}(x), \xi_{\alpha,j}(y)\rangle\right| : d_\alpha(x,y) \geq S,\ x, y \in U^R_{\alpha,j}\right\} = 0.$

For each $\alpha \in I$, extend $\xi_{\alpha,j}$ to all of X_α by setting $\xi_{\alpha,j}(x) = 0$ if $x \in X_\alpha \setminus U^R_{\alpha,j}$. Then we can define the map $\eta_\alpha : X_\alpha \to H_\alpha = \oplus_{j \in J_\alpha} H_{\alpha,j}$, $\eta_\alpha(x) = \left(\eta_{\alpha,j}(x)\right)_{j \in J_\alpha}$, by setting

$$\eta_{\alpha,j}(x) = \varphi_{\alpha,j}(x)^{1/2} \xi_{\alpha,j}(x).$$

It now follows from [7, proof of Theorem 3.2] that

(i′) $\|\eta_\alpha(x)\| = 1$, for all $x \in X_\alpha$ and $\alpha \in I$;
(ii′) $\forall \alpha \in I$, $\forall x, y \in X_\alpha$, $d_\alpha(x,y) \leq R \Rightarrow \|\eta_\alpha(x) - \eta_\alpha(y)\| \leq \varepsilon$;
(iii′) $\lim\limits_{S \to \infty} \sup\limits_{\alpha \in I} \sup\left\{\left|\langle \eta_\alpha(x), \eta_\alpha(y)\rangle\right| : d_\alpha(x,y) \geq S,\ x, y \in X_\alpha\right\} = 0.$

Thus, by Proposition 9.22, $\mathcal{X} = \{X_\alpha \mid \alpha \in I\}$ is in $\mathfrak{H}$. □

Remark In [12, Theorem 4.6], Guentner, Tessera and Yu proved that the collection of *exact*[2] metric families, $\mathfrak{E}$, is closed under weak decomposition. Thus, since $\mathfrak{B}$ is contained in $\mathfrak{E}$, every metric family with weak finite decomposition complexity is also in $\mathfrak{E}$. A straightforward generalization of [7, Proposition 2.10(c)] shows that an exact metric family is coarsely embeddable into Hilbert space. Therefore, we have the following sequence of inclusions of collections of metric families, each of which is stable under decomposition:

$$w\mathfrak{D} \subset \mathfrak{E} \subset \mathfrak{H}.$$

9.3 Weak Hyperbolic Dimension

In this section we prove that a metric space with finite *hyperbolic dimension*, and more generally one with finite *weak hyperbolic dimension*, has weak finite decomposition complexity (Theorem 9.29). Buyalo and Schroeder introduced the hyperbolic dimension of a metric space (Definition 9.25) to study the quasi-isometric embedding properties of negatively curved spaces (see [4] for an exposition). The related notion of weak hyperbolic dimension was introduced by Cappadocia in his PhD thesis [5].

Definition 9.24 Let N be a positive integer and $R > 0$. A subset $Y \subset X$ of a metric space (X, d) is (N, R)*-large scale doubling* if for every $x \in X$ and every $r \geq R$, the intersection of Y with a ball in X with radius $2r$ centered at x can be covered with N balls of radius r with centers in X.

A metric family $\mathcal{Y}$ of subsets of X is *large scale doubling*[3] if there exists (N, R) such that each $Y \in \mathcal{Y}$ is (N, R)-large scale doubling and every finite union of elements of $\mathcal{Y}$ is (N, R')-large scale doubling, where possibly $R' > R$ and R' could depend on the particular finite union.

Hyperbolic dimension is analogous to asymptotic dimension with the role of bounded metric families replaced by large scale doubling metric families.

[2] *Exactness* of a metric space is a coarse invariant related to the notion of *Property A*. Specifically, a metric space with Property A is exact, and an exact metric space with bounded geometry has Property A [6].

[3] Some authors call such a collection of subsets *uniformly large scale doubling*.

Definition 9.25 Let n be a non-negative integer. Let $\mathfrak{L}$ be the collection of large scale doubling metric families. A metric space (X, d) has *hyperbolic dimension* at most n, denoted $\operatorname{hyperdim}(X) \leq n$, if $\{X\}$ is n-decomposable over $\mathfrak{L}$. We say $\operatorname{hyperdim}(X) = n$ if n is the smallest non-negative integer for which $\operatorname{hyperdim}(X) \leq n$. If no such integer exists then, by convention, $\operatorname{hyperdim}(X) = \infty$.

Since a bounded metric family is large scale doubling, we have that $\operatorname{hyperdim}(X) \leq \operatorname{asdim}(X)$. If X is a large scale doubling metric space (for example, $\mathbb{R}^n$ with the Euclidean metric), then $\operatorname{hyperdim}(X) = 0$. Buyalo and Schroeder showed that $\operatorname{hyperdim}(\mathbb{H}^n) = n$, where $\mathbb{H}^n$ is n-dimensional hyperbolic space, $n \geq 2$. Chris Cappadocia introduced the *weak hyperbolic dimension* of a metric space in his PhD thesis [5]. In Cappadocia's theory, large scale doubling metric families are replaced by *weakly large scale doubling*[4] metric families, dropping the condition on finite unions appearing in Definition 9.25. That is, a metric family $\mathcal{Y}$ of subsets of a metric space X is called *weakly large scale doubling* if there exists (N, R) such that each $Y \in \mathcal{Y}$ is (N, R)-large scale doubling.

Definition 9.26 ([5]) Let $w\mathfrak{L}$ be the collection of weakly large scale doubling metric families. A metric space (X, d) has *weak hyperbolic dimension* at most n if $\{X\}$ is n-decomposable over $w\mathfrak{L}$. We denote this property by $\text{w-hyperdim}(X) \leq n$, and write $\text{w-hyperdim}(X) = n$ if n is the smallest non-negative integer for which $\text{w-hyperdim}(X) \leq n$. If no such integer exists then, by convention, $\text{w-hyperdim}(X) = \infty$.

Since $\mathfrak{L} \subset w\mathfrak{L}$, we have $\text{w-hyperdim}(X) \leq \operatorname{hyperdim}(X) \leq \operatorname{asdim}(X)$.

We say that a metric space is *(N, R)-large scale doubling* if it is (N, R)-large scale doubling as a subset of itself (see Definition 9.24).

Lemma 9.27 *Let $U \subset X$ be an (N, R)-large scale doubling subset of a metric space (X, d_X). Then (U, d_U) is $(N^2, 2R)$-large scale doubling, where d_U is the subspace metric induced by d_X.*

Proof Let $x \in U$ and $r \geq 2R$. Since U is an (N, R)-large scale doubling subset of X, there are points $x_1, \ldots, x_{N^2} \in X$ such that $B_{2r}(x) \subset \bigcup_{i=1}^{N^2} B_{r/2}(x_i)$. Let J be the set of indices, i, for which $B_{r/2}(x_i) \cap U$ is non-empty. For each $i \in J$ choose $u_i \in B_{r/2}(x_i) \cap U$. Since $B_{r/2}(x_i) \subset B_r(u_i)$ for $i \in J$, we have that $B_{2r}(x) \cap U \subset \bigcup_{i \in J} B_r(u_i) \cap U$. □

A subset $A \subset X$ of a metric space (X, d_X) is said to be *L-separated,*

[4] Cappadocia uses the terminology *uniformly weakly large scale doubling.*

where $L > 0$, if $d_X(u, v) \geq L$ for all $u, v \in A$ with $u \neq v$. We say that a metric space (X, d_X) is *N-doubling*, where N is a positive integer, if it is (N, R)-large scale doubling for all $R > 0$, that is, doubling at all scales with doubling constant N.

We are now able to prove the key fact needed to establish Theorem 9.29.

Proposition 9.28 *Let $\mathcal{X} = \{(X_\alpha, d_\alpha) \mid \alpha \in I\}$ be a metric family such that there exists (N, R) with the property that each (X_α, d_α) is (N, R)-large scale doubling. Then there exists a positive integer M, depending only on N, such that* $\operatorname{asdim}(\mathcal{X}) \leq M$.

Proof Let $\lambda > 0$ be given. Let $r = \max(\lambda, R)$. For each $\alpha \in I$, choose a maximal $2r$-separated set $Z_\alpha \subset X_\alpha$. Then $\mathcal{U}_\alpha = \{B_{4r}(x) \mid x \in Z_\alpha\}$ is a cover of X_α. Note that the Lebesgue number of $\mathcal{U}_\alpha$ satisfies $L(\mathcal{U}_\alpha) \geq \lambda$, for each $\alpha \in I$.

Let ℓ be a positive integer and assume that $y \in B_{4r}(x_1) \cap \cdots \cap B_{4r}(x_\ell)$, where $x_1, \ldots, x_\ell \in Z_\alpha$ are distinct. Note that $\{x_1, \ldots, x_\ell\} \subset B_{8r}(x_1)$. By Lemma 9.27, $B_{8r}(x_1) \cap Z_\alpha$ can be covered by N^2 balls of radius $4r$ with centers in Z_α and, in turn, each of these balls can be covered by N^2 balls of radius $2r$ with centers in Z_α. For each $z \in Z_\alpha$, $B_{2r}(z) \cap Z_\alpha = \{z\}$ because Z_α is $2r$-separated. It follows that $B_{8r}(x_1) \cap Z_\alpha$ contains at most N^4 points, and so $\ell \leq N^4$. Hence, for each $\alpha \in I$, the multiplicity of the cover $\mathcal{U}_\alpha$ is at most N^4. Since $\cup_{\alpha \in I} \mathcal{U}_\alpha$ is a bounded metric family, Proposition 9.18 implies that $\mathcal{X}$ is $(N^4 - 1)$-decomposable over $\mathfrak{B}$ (the collection of bounded metric families). In other words, $\operatorname{asdim}(\mathcal{X}) \leq N^4 - 1$. □

Combining Definition 9.26 and Proposition 9.28 yields the following theorem.

Theorem 9.29 *A metric space X with finite weak hyperbolic dimension has weak finite decomposition complexity. If X has weak hyperbolic dimension at most 1, then X has (strong) finite decomposition complexity.*

Proof If $\text{w-hyperdim}(X) \leq n$, then, applying Proposition 9.28, X is n-decomposable over $\mathfrak{A}$ (the collection of metric families with finite asymptotic dimension). Therefore by (9.1), X has weak FDC, and if $n \leq 1$, then X has FDC. □

Since $\text{w-hyperdim}(X) \leq \text{hyperdim}(X)$, we also get the following corollary.

Corollary 9.30 *A metric space X with finite hyperbolic dimension has weak finite decomposition complexity. If X has hyperbolic dimension at most 1, then X has (strong) finite decomposition complexity.* □

9.4 Some Open Questions

In this section we discuss some open problems involving decomposition complexity.

There are some interesting finitely generated groups for which the FDC (or weak FDC) condition is unknown.

Question *Consider the following groups:*

1 Grigorchuk's group of intermediate growth,
2 Thompson's group

$$F = \langle A, B \mid [AB^{-1}, A^{-1}BA] = [AB^{-1}, A^{-2}BA^2] = 1 \rangle,$$

3 $\mathrm{Out}(F_n)$, *the outer automorphism group of a free group F_n of rank* $n \geq 3$.

For which of these groups, if any, does the FDC (or weak FDC) condition hold?

Grigorchuk's group and Thompson's group F are known to have infinite asymptotic dimension. Grigorchuk's group is amenable and therefore has Yu's *Property A*, a condition that implies coarse embeddability into Hilbert space (see [12, Section 4] for a discussion of Property A). A group with weak FDC has Property A, but the reverse implication is unknown.

Question *Does Property A for a countable group imply weak FDC?*

Osajda gave an example of a finitely generated group that is coarsely embeddable into Hilbert space yet does not have Property A [14]. Thus, Osajda's example is a group in the collection $\mathfrak{H}$ of metric families that are coarsely embeddable into Hilbert space, but not in the collection $\mathfrak{E}$ of exact metric families.

Question *Are there interesting collections of metric families, stable under (weak or strong) decomposition, that lie strictly in between $\mathfrak{E}$ and $\mathfrak{H}$?*

As pointed out to us by the referee, while our proof of Theorem 9.23 is very specific to Hilbert spaces, it is natural to try to generalize it to more general classes of Banach spaces. That is:

Question *Is there an interesting class of Banach spaces that is stable under weak decomposition?*

The mapping class group of a surface has finite asymptotic dimension [3], and, by analogy, one surmises that $\mathrm{Out}(F_n)$ may also have finite asymptotic dimension and hence FDC. Although a proof that the asymptotic dimension of $\mathrm{Out}(F_n)$ is finite has so far been elusive, perhaps the less restrictive, yet geometrically consequential (see the discussion in Section 9.1) weak FDC condition might be easier to demonstrate.

Question *Which, if any, of the groups: Grigorchuk's group, Thompson's group F and $\mathrm{Out}(F_n)$, $n \geq 3$, have finite weak hyperbolic dimension?*

None of these groups are large scale doubling as metric spaces and so their weak hyperbolic dimension is at least 1. Note that by Theorem 9.29, any group on this list that has finite weak hyperbolic dimension must have weak FDC.

Question *Does a space with finite weak hyperbolic dimension have FDC?*

This question may be more tractable than the general question of whether weak FDC implies FDC (see [12, Question 2.2.6]).

Acknowledgments

A. N. was partially supported by a grant from the Natural Sciences and Engineering Research Council of Canada. D. R. was partially supported by a grant from the Simons Foundation, #229577.

References

[1] Bell, G. and Dranishnikov, A. 2004. On asymptotic dimension of groups acting on trees. *Geom. Dedicata*, **103**, 89–101.

[2] Bell, G. and Dranishnikov, A. 2011. Asymptotic dimension in Bedlewo. *Topology Proc.*, **38**, 209–236.

[3] Bestvina, Mladen, Bromberg, Ken and Fujiwara, Koji. 2015. Constructing group actions on quasi-trees and applications to mapping class groups. *Publ. Math. Inst. Hautes Études Sci.*, **122**, 1–64.

[4] Buyalo, Sergei and Schroeder, Viktor. 2007. *Elements of asymptotic geometry.* EMS Monographs in Mathematics. European Mathematical Society (EMS).

[5] Cappadocia, C. 2014. *Large scale dimension theory of metric spaces.* Ph.D. thesis, McMaster University.

[6] Dadarlat, Marius and Guentner, Erik. 2003. Constructions preserving Hilbert space uniform embeddability of discrete groups. *Trans. Amer. Math. Soc.*, **355**(8), 3253–3275.

[7] Dadarlat, Marius and Guentner, Erik. 2007. Uniform embeddability of relatively hyperbolic groups. *J. Reine Angew. Math.*, **612**, 1–15.

[8] Goldfarb, B. 2013. *Weak coherence of groups and finite decomposition complexity.* To appear in Int. Math. Res. Not. Preprint available at `http://arxiv.org/abs/1307.5345`.

[9] Grave, Bernd. 2006. Asymptotic dimension of coarse spaces. *New York J. Math.*, **12**, 249–256.

[10] Gromov, M. 1993. Asymptotic invariants of infinite groups. Pages 1–295 of: *Geometric group theory, Vol. 2 (Sussex, 1991).* London Math. Soc. Lecture Note Ser., vol. 182. Cambridge University Press.

[11] Guentner, Erik, Tessera, Romain and Yu, Guoliang. 2012. A notion of geometric complexity and its application to topological rigidity. *Invent. Math.*, **189**(2), 315–357.

[12] Guentner, Erik, Tessera, Romain and Yu, Guoliang. 2013. Discrete groups with finite decomposition complexity. *Groups Geom. Dyn.*, **7**(2), 377–402.

[13] Kasprowski, Daniel. 2015. On the K-theory of groups with finite decomposition complexity. *Proc. Lond. Math. Soc. (3)*, **110**(3), 565–592.

[14] Osajda, D. 2014. *Small cancellation labellings of some infinite graphs and applications.* Preprint available at `http://arxiv.org/abs/1406.5015`.

[15] Ramras, Daniel A., Tessera, Romain and Yu, Guoliang. 2014. Finite decomposition complexity and the integral Novikov conjecture for higher algebraic K-theory. *J. Reine Angew. Math.*, **694**, 129–178.

[16] Wu, Yan and Chen, Xiaoman. 2011. On finite decomposition complexity of Thompson group. *J. Funct. Anal.*, **261**(4), 981–998.

[17] Yu, Guoliang. 1998. The Novikov conjecture for groups with finite asymptotic dimension. *Ann. of Math. (2)*, **147**(2), 325–355.

10

Some Remarks on the Covering Groups of a Topological Group

Dongwen Qi

Abstract

We discuss the connection between Chevalley's definition of a covering space and the usual definition given in an introductory topology course. Then we indicate how some theorems about the covering groups of a topological group can be proved from the global point of view, without using local isomorphisms between topological groups.

10.1 Covering Spaces

In a beginning topology course (e.g. [2]), a covering space of a topological space is defined as follows.

Definition 10.1 For a topological space B (often abbreviated as a space B in this chapter), a *covering space* (E, p) of B is a pair consisting of a space E and a continuous surjective mapping $p : E \to B$, such that each point of B has an *open* neighborhood V, whose inverse image $p^{-1}(V)$ is a union of disjoint open sets U_α in E, with the property that $p|_{U_\alpha} : U_\alpha \to V$ is a homeomorphism.

However, the idea of covering space is useful in algebraic topology only when some more conditions are imposed, for example, when the spaces involved are assumed to be connected and locally path-connected. In this case, the lifting property can be proved with the help of considering fundamental groups of the corresponding spaces. Also, the existence of a simply connected covering space (called a *universal covering*) is proved for a connected, locally path-connected, and semilocally simply connected space B. The universal covering E is constructed by using path homotopy classes in B.

Chevalley's monograph *Theory of Lie Groups I* [1] revolutionized the idea of viewing Lie groups as global objects. It is well known that Lie groups can admit non-trivial covering groups. However, Chevalley preferred to develop the theory of covering spaces without using paths or local path-connectedness. In furtherance of this effort, he formulated the following definition of a covering space, where connectedness and local connectedness of the space E are assumed, for the purpose of proving the lifting property without using paths. The connectedness of some classical Lie groups had been proved through a standard procedure of considering homogeneous spaces (and induction). We assume that all the spaces are *Hausdorff* in the following discussions.

Definition 10.2 ([1], page 40) For a continuous mapping p of a space E into a space B, a subset B_1 of B is said to be *evenly covered* by E (with respect to p) if $p^{-1}(B_1)$ is non-empty, and every (connected) component of $p^{-1}(B_1)$ is mapped homeomorphically onto B_1 by the mapping p.

Definition 10.3 ([1], page 40) For a space B, a *covering space* (E, p) of B is a pair formed by a connected and locally connected space E and a continuous mapping p of E onto B which has the following property: each point of B has a neighborhood which is *evenly covered* by E (with respect to p).

Here a neighborhood of a point b in space B is understood to be a set N such that there exists an open set U such that $b \in U \subset N$; N need not be open itself.

It is worth noting that *in a locally connected space, every component of an open set is an open set.*

Now we show that Chevalley's definition of a covering space satisfies the usual requirements of a covering space defined in a typical topology course.

Proof Given a point $b \in B$, let N_1 be a neighborhood of b that is evenly covered by p. Choose an open set U_1 of B such that $b \in U_1 \subset N_1$, and $a \in p^{-1}(\{b\}) \subset E$. Since E is locally connected, continuity of p enables us to find a connected open neighborhood U of a in E such that $p(U) \subset U_1 \subset N_1$.

Since N_1 is evenly covered by p, the connected neighborhood U of a is contained in one of the connected components of $p^{-1}(N_1)$, which is mapped homeomorphically onto N_1 in the subspace topology. Thus

$p(U)$ is relatively open in N_1, i.e., being the intersection of an open set W_1 (of B) with N_1. Note also that $p(U) \subset U_1 \subset N_1$, and U_1 is open in B, we see that $p(U) = W_1 \cap U_1$ is indeed an open set of B.

The preceding discussion implies that B is locally connected, and is connected (since $B = p(E)$). Now let $V_1 \subset N_1$ be a connected open neighborhood of a. Each connected component O_α of $p^{-1}(V_1)$ is contained in a component M_α of $p^{-1}(N_1)$. Since E is locally connected and $p^{-1}(V_1)$ is open, O_α is open in E. In each component M_β of $p^{-1}(N_1)$, there is a subset $O'_\beta \subset M_\beta$ such that $p(O'_\beta) = V_1$. It follows that the sets of indices $\{\alpha\}$ and $\{\beta\}$ are the same, and $O_\alpha = O'_\alpha$. This shows that Chevalley's definition of a covering space satisfies the usual requirements described in a topology course, when E *is assumed to be connected and locally connected.*

□

10.2 Covering Groups of a Topological Group

A topological group G is called a *covering group* of a topological group H if there is a continuous homomorphism $f : G \to H$, such that the pair (G, f) is a covering space of H.

The following statement is proved in [1] (page 53) using the lifting property related to a universal covering.

Proposition 10.4 *Assume that a topological group H has a simply connected covering space (G, f). It is then possible to define a multiplication in G that turns the space G into a topological group and the covering space (G, f) into a covering group.*

Then Chevalley went on to prove the following statement by considering extensions of local isomorphisms.

Proposition 10.5 *If a topological group H admits a simply connected covering group (G, f), this covering group is unique up to isomorphism; i.e., if (G', f') is another simply connected covering group of H, then there exists an isomorphism θ of the topological group G with G' such that $f = f' \circ \theta$.*

We now show that Proposition 10.5 can be proved by simply applying the lifting property related to a universal covering. The lifting property is proved in [1] (page 50) under the assumption that G (or E) is connected, locally connected, and simply connected; or we may replace local connectedness by local path-connectedness (which is the case of Lie groups) and apply the typical path-lifting arguments provided in a topology course (see [2]).

Proof Let e and e' be the identity elements in G and G', respectively. Since both G and G' are simply connected, it follows from the lifting property that there is a unique continuous mapping $\theta : G \to G'$, such that $\theta(e) = e'$, and $f = f' \circ \theta$. A continuous mapping $\theta' : G' \to G$ is determined similarly that satisfies $\theta'(e') = e$ and $f' = f \circ \theta'$.

Fix an $x \in G$, consider two mappings m_x and n_x from G to G', where $m_x(y) = \theta(xy)$, and $n_x(y) = \theta(x)\theta(y)$, for $y \in G$. It is clear that both mappings are continuous, and $m_x(e) = \theta(x) = n_x(e)$. Note that $f'(m_x(y)) = f'(\theta(xy)) = f(xy) = f(x)f(y)$, and $f'(n_x(y)) = f'(\theta(x)\theta(y)) = f'(\theta(x))f'(\theta(y)) = f(x)f(y)$, since both f and f' are group homomorphisms. Uniqueness of lifting implies $\theta(xy) = \theta(x)\theta(y)$, for $x, y \in G$, and $\theta' \circ \theta = \mathrm{id}_G$, and $\theta \circ \theta' = \mathrm{id}_{G'}$, because $f \circ (\theta' \circ \theta) = (f \circ \theta') \circ \theta = f' \circ \theta = f$, etc. Thus Proposition 10.5 is proved without incorporating local isomorphisms among the three groups. □

For a space B which has a simply connected covering space (E, p), the group of *deck transformations* of B, that is, the group of homeomorphisms $\theta : E \to E$ such that $p \circ \theta = p$, is isomorphic to the *fundamental group* of B.

The following statement ([1], page 59) is important when discussing the fundamental groups of classical Lie groups.

Proposition 10.6 *Let G be a connected and locally connected topological group, and let H be a closed locally connected subgroup of G. Assume that G/H is simply connected and that G and H are semilocally simply connected. Then the fundamental group of G is isomorphic to a factor group of the fundamental group of H.*

The conditions in Proposition 10.6 imply that H is connected, and both G and H admit simply connected covering groups, respectively. Here we give a proof of the above proposition using lifting properties. Again, we do not need to consider extending local isomorphisms.

Proof Let $(\widetilde{G}, p)$ and $(\widetilde{H}, q)$ denote simply connected covering groups of G and H respectively. Write $K = p^{-1}(H)$. Then K is a closed subgroup of $\widetilde{G}$. By considering coset correspondences and the corresponding quotient topologies, it is shown (in [1], page 60) that $\widetilde{G}/K$ and G/H are homeomorphic. It is not difficult to verify that K is locally connected, and hence is connected. The fact that $p_1 = p|_K : K \to H$ is a covering map follows from a standard argument in topology.

Now we proceed to show that the fundamental group of G is isomorphic to a factor group of the fundamental group of H.

Let e_1 and e_2 be the identity elements in $\widetilde{G}$ and $\widetilde{H}$ respectively. Because of the lifting property of universal covering spaces, the group of deck transformations of G is isomorphic to the kernel $\ker(p)$ of $p : \widetilde{G} \to G$ ([1], page 54). The fact that there is a continuous map $\phi : \widetilde{H} \to K$ such that $q = p_1 \circ \phi$ and $\phi(e_2) = e_1$ also follows from the lifting property. The basic theory of covering spaces ([2], page 485) implies that $\phi : \widetilde{H} \to K$ is a covering map, and hence is surjective. Since $p_1(\phi(xy)) = q(xy) = q(x)q(y)$, $p_1(\phi(x)\phi(y)) = p_1(\phi(x))p_1(\phi(y)) = q(x)q(y)$, for $x, y \in \widetilde{H}$, and $\phi(xe_2) = \phi(x) = \phi(x)e_1 = \phi(x)\phi(e_2)$, the lifting property implies that $\phi(xy) = \phi(x)\phi(y)$. So $\phi : \widetilde{H} \to K$ is a group homomorphism.

Since $q = p_1 \circ \phi$, $\phi(\ker(q)) \subset \ker(p_1) = \ker(p) \subset K$. For any $z \in \ker(p)$, there is an $x \in \widetilde{H}$, $\phi(x) = z$. Thus $q(x) = p_1(\phi(x)) = p_1(z)$, which is the identity element in $H \subset G$. It follows that $\phi|_{\ker(q)} : \ker(q) \to \ker(p)$ is surjective. Hence the group of deck transformations of $\widetilde{G} \to G$ is isomorphic to a factor group of the group of deck transformations of $\widetilde{H} \to H$.

□

Acknowledgement

The author would like to thank the referee for the helpful suggestions.

References

[1] Chevalley, Claude. 1946. *Theory of Lie Groups. I.* Princeton Mathematical Series, vol. 8. Princeton University Press.

[2] Munkres, James R. 2000. *Topology (2nd edition).* Prentice-Hall, Inc.

11

The Σ-invariants of Thompson's group F via Morse Theory

Stefan Witzel and Matthew C. B. Zaremsky

Abstract

Bieri–Geoghegan–Kochloukova computed the BNSR-invariants $\Sigma^m(F)$ of Thompson's group F for all m. We recompute these using entirely geometric techniques, making use of the Stein–Farley CAT(0) cube complex X on which F acts.

11.1 Introduction

In [5], Bieri, Geoghegan and Kochloukova computed $\Sigma^m(F)$ for all m. Here F is Thompson's group, and Σ^m is the mth *Bieri–Neumann–Strebel–Renz (BNSR) invariant*. This is a topological invariant of a group of type F_m [4, 3]. The proof in [5] makes use of various algebraic facts about F, for instance that it contains no non-abelian free subgroups, and that it is isomorphic to an ascending HNN-extension of an isomorphic copy of itself, and applies tools specific to such groups to compute all the $\Sigma^m(F)$.

In this chapter we consider the free action of F by isometries on a proper CAT(0) cube complex X, the *Stein–Farley complex*. We apply a version of Bestvina–Brady Morse theory to this space, and recompute the invariants $\Sigma^m(F)$. The crucial work to do, using this approach, is to analyze the homotopy type of *ascending links* of vertices in X and *superlevel sets* in X with respect to various *character height functions*. Our proof is purely geometric and self-contained, and does not require the aforementioned algebraic facts about F. It is possible that this approach could be useful in determining the BNSR-invariants of other interesting groups.

The abelianization of Thompson's group F is free abelian of rank 2. A basis of $\mathrm{Hom}(F, \mathbb{R}) \cong \mathbb{R}^2$ is given by two homomorphisms usually denoted χ_0 and χ_1. The main result of [5] is the following.

Theorem A *Write a non-trivial character $\chi\colon F \to \mathbb{R}$ as $\chi = a\chi_0 + b\chi_1$ (with $(a,b) \neq (0,0)$). Then $[\chi]$ is in $\Sigma^1(F)$ unless $a > 0$ and $b = 0$, or $b > 0$ and $a = 0$. Moreover, $[\chi]$ is in $\Sigma^2(F) = \Sigma^\infty(F)$ unless $a \geq 0$ and $b \geq 0$.*

Here $[\chi]$ denotes the equivalence class of χ up to positive scaling.

In Section 11.2 we recall the invariants $\Sigma^m(G)$, and set up the Morse theory tools we will use. In Section 11.3 we recall Thompson's group F and the Stein–Farley complex X, and discuss characters of F and height functions on X. Section 11.4 is devoted to combinatorially modeling vertex links in X, which is a helpful tool in the computations of the $\Sigma^m(F)$. Theorem A is proven in stages in Sections 11.5 through 11.7.

11.2 The Invariants

A *character* of a group G is a homomorphism $\chi\colon G \to \mathbb{R}$. If the image is infinite cyclic, the character is *discrete*. If G is finitely generated, we can take $\mathrm{Hom}(G, \mathbb{R}) \cong \mathbb{R}^d$ and mod out scaling by positive real numbers to get $S(G) := S^{d-1}$, called the *character sphere*. Here d is the rank of the abelianization of G. Now, the definition of the *Bieri–Neumann–Strebel (BNS) invariant* $\Sigma^1(G)$ for G finitely generated (first introduced in [4]) is the subset of $S(G)$ defined by:

$$\Sigma^1(G) := \{[\chi] \in S(G) \mid \Gamma_{0\leq\chi} \text{ is connected}\}.$$

Here Γ is the Cayley graph of G with respect to some finite generating set, and $\Gamma_{0\leq\chi}$ is the full subgraph spanned by those vertices g with $0 \leq \chi(g)$.

The higher *Bieri–Neumann–Strebel–Renz (BNSR) invariants* $\Sigma^m(G)$ ($m \in \mathbb{N} \cup \{\infty\}$), introduced in [3], are defined somewhat analogously. Let G be of type F_m and let χ be a character of G. Let Γ^m be the result of equivariantly gluing cells, up to dimension m, to the Cayley graph Γ, in a G-cocompact way to produce an $(m-1)$-connected space (that this is possible is more or less the definition of being of type F_m). For $t \in \mathbb{R}$ let $\Gamma^m_{t\leq\chi}$ be the full subcomplex spanned by those g with $t \leq \chi(g)$. Then $[\chi] \in \Sigma^m(G)$ by definition if the filtration $(\Gamma^m_{t\leq\chi})_{t\in\mathbb{R}}$ is essentially $(m-1)$-connected.

Recall that a space Y is $(m-1)$-*connected* if every continuous map $S^{k-1} \to Y$ is homotopic to a constant map for $k \leq m$. A filtration $(Y_t)_{t\in\mathbb{R}}$ (with $Y_t \supseteq Y_{t+1}$) is *essentially* $(m-1)$-*connected* if for every t there is a $t' \leq t$ such that for all $k \leq m$, every continuous map $S^{k-1} \to Y_t$ is homotopic in $Y_{t'}$ to a constant map. Recall also that a pair (Y, Y_0) is $(m-1)$-*connected* if for all $k \leq m$, every map $(D^{k-1}, S^{k-2}) \to (Y, Y_0)$ is homotopic relative Y_0 to a map with image in Y_0.

In the realm of finiteness properties, the main application of the BNSR-invariants is the following, see [3, Theorem 5.1] and [5, Theorem 1.1].

Theorem 11.1 *[5, Theorem 1.1] Let G be a group of type* F_m *and N a normal subgroup of G with G/N abelian. Then N is of type* F_m *if and only if for every $\chi \in \mathrm{Hom}(G, \mathbb{R})$ with $\chi(N) = 0$ we have $[\chi] \in \Sigma^m(G)$.*

As an example, if χ is discrete then $\ker(\chi)$ is of type F_m if and only if $[\pm\chi] \in \Sigma^m(G)$.

It turns out that to compute $\Sigma^m(G)$, one can use spaces and filtrations other than just Γ^m and $(\Gamma^m_{t\leq\chi})_{t\in\mathbb{R}}$.

Proposition 11.2 *[9, Definition 8.1] Let G be of type* F_m*, acting cellularly on an $(m-1)$-connected CW complex Y. Suppose the stabilizer of any k-cell is of type* F_{m-k}*, and that the action of G on $Y^{(m)}$ is cocompact. For any non-trivial character $\chi \in \mathrm{Hom}(G, \mathbb{R})$, there is a* character height function*, denoted h_χ, i.e., a continuous map $h_\chi \colon Y \to \mathbb{R}$, such that $h_\chi(gy) = \chi(g) + h_\chi(y)$ for all $y \in Y$ and $g \in G$. Then $[\chi] \in \Sigma^m(G)$ if and only if the filtration $(h_\chi^{-1}([t, \infty)))_{t\in\mathbb{R}}$ is essentially $(m-1)$-connected.*

It is a fact that this is independent of Y and h_χ. In our case G can be naturally regarded as a subset of Y and we will therefore write χ to denote both the character and the character height function.

The filtration of Y that we will actually use in practice is $(Y_{t\leq\chi})_{t\in\mathbb{R}}$, where $Y_{t\leq\chi}$ is defined to be the full subcomplex of Y supported on those vertices y with $t \leq \chi(y)$. Our working definition of $\Sigma^m(G)$ will be the following.

Corollary 11.3 (Working definition) *Let G, Y and χ be as in Proposition 11.2. Then $[\chi] \in \Sigma^m(G)$ if and only if the filtration $(Y_{t\leq\chi})_{t\in\mathbb{R}}$ is essentially $(m-1)$-connected.*

Proof We need to show that $(Y_{t\leq\chi})_{t\in\mathbb{R}}$ is essentially $(m-1)$-connected if and only if $(h_\chi^{-1}([t, \infty)))_{t\in\mathbb{R}}$ is. First note that this is really a statement about m-skeleta. Then since m-skeleta are G-cocompact, there exists a uniform bound D such that for any two points y, y' of Y sharing a k-cell ($k \leq m$), we have $|\chi(y) - \chi(y')| \leq D$. Now the result is immediate, since $Y_{t\leq\chi}^{(m)} \subseteq \chi^{-1}([t-D,\infty))^{(m)}$ and $\chi^{-1}([t,\infty))^{(m)} \subseteq Y_{t-D\leq\chi}^{(m)}$. □

Remark Bieri and Geoghegan generalized the BNSR-invariants in [2], to a family of invariants $\Sigma^m(\rho)$ defined for any (sufficiently nice) action ρ of a group on a CAT(0) space. The classical BNSR-invariants agree with the case where ρ is the action of G on $\mathrm{Hom}(G, \mathbb{R})$. See also [12, Section 18.3].

11.2.1 Morse Theory

The criterion Corollary 11.3 is particularly useful in the situation where Y is an affine cell complex and χ is affine on cells. One can then make use of a version of Bestvina–Brady Morse theory and study relative connectivity locally in terms of ascending/descending links.

By an *affine cell complex* we mean a complex that is obtained by gluing together euclidean polytopes. More precisely, Y is an affine cell complex if it is the quotient $\hat{Y}/\sim$ of a disjoint union $\hat{Y} = \bigcup_\lambda C_\lambda$ of euclidean polytopes modulo an equivalence relation $\sim$ such that every polytope is mapped injectively to Y, and such that if two faces of polytopes have a (relative) interior point identified then their entire (relative) interior is isometrically identified (see [7, Definition I.7.37] for a more general definition). In particular, every cell (meaning every image of some polytope in $\hat{Y}$) carries an affine structure. The link $\mathrm{lk}_Y v$ of a vertex v of Y consists of directions issuing at the vertex. It naturally carries the structure of a spherical cell complex, whose closed cells consist of directions that point into closed cells of Y.

Definition 11.4 (Morse function) The most general kind of *Morse function* on Y that we will be using is a map $(h, s)\colon Y \to \mathbb{R} \times \mathbb{R}$ such that both h and s are affine on cells. The codomain is ordered lexicographically, and the conditions for (h, s) to be a Morse function are the following: the function s takes only finitely many values on vertices of Y, and there is an $\varepsilon > 0$ such that every pair of adjacent vertices v and w either satisfy $|h(v) - h(w)| \geq \varepsilon$, or else $h(v) = h(w)$ and $s(v) \neq s(w)$.

As an example, if h is discrete and s is constant, we recover the notion of "Morse function" from [1]. We think of a Morse function as assigning a *height* $h(v)$ to each vertex v. The secondary height function s "breaks ties" for adjacent vertices of the same height. We will speak of $(h,s)(v)$ as the *refined height* of v. Note that since h is affine, the set of points of a cell σ on which h attains its maximum is a face $\bar{\sigma}$. Since s is affine as well, the set of points of $\bar{\sigma}$ on which s attains its maximum is a face $\hat{\sigma}$. If $\hat{\sigma}$ were to have two adjacent vertices, these vertices would have the same refined height, and (h,s) would not be a Morse function. This shows that every cell has a unique vertex of maximal refined height and (by symmetry) a unique vertex of minimal refined height.

The *ascending star* $\operatorname{star}^{(h,s)\uparrow} v$ of a vertex v (with respect to (h,s)) is the subcomplex of $\operatorname{star} v$ consisting of cells σ such that v is the vertex of minimal refined height in σ. The *ascending link* $\operatorname{lk}^{(h,s)\uparrow} v$ of v is the link of v in $\operatorname{star}^{(h,s)\uparrow} v$. The *descending star* and the *descending link* are defined analogously. A consequence of h and s being affine is the following.

Observation 11.5 *Ascending and descending links are full subcomplexes.* □

We use notation like $Y_{p\le h\le q}$ to denote the full subcomplex of Y supported on the vertices v with $p \le h(v) \le q$ (this is the union of the closed cells all of whose vertices lie within the bounds). An important tool we will use is the following.

Lemma 11.6 (Morse Lemma) *Let $p,q,r \in \mathbb{R} \cup \{\pm\infty\}$ be such that $p \le q \le r$. If for every vertex $v \in Y_{q<h\le r}$ the descending link $\operatorname{lk}^{(h,s)\downarrow}_{Y_{p\le h}} v$ is $(k-1)$-connected then the pair $(Y_{p\le h\le r}, Y_{p\le h\le q})$ is k-connected. If for every vertex $v \in Y_{p\le h<q}$ the ascending link $\operatorname{lk}^{(h,s)\uparrow}_{Y_{h\le r}} v$ is $(k-1)$-connected then the pair $(Y_{p\le h\le r}, Y_{q\le h\le r})$ is k-connected.*

Proof The second statement is like the first with (h,s) replaced by $-(h,s)$, so we only prove the first. Using induction (and compactness of spheres in case $r=\infty$) we may assume that $r-q \le \varepsilon$ (where $\varepsilon > 0$ is as in Definition 11.4). By compactness of spheres, it suffices to show that there exists a well order $\preceq$ on the vertices of $Y_{q<h\le r}$ such that the pair

$$(S_{\preceq v}, S_{\prec v}) := \left(Y_{p\le h\le q} \cup \bigcup_{w\preceq v} \operatorname{star}^{(h,s)\downarrow}_{Y_{p\le h}} w,\ Y_{p\le h\le q} \cup \bigcup_{w\prec v} \operatorname{star}^{(h,s)\downarrow}_{Y_{p\le h}} w \right)$$

is k-connected for every vertex $v \in Y_{q<h\le r}$. To this end, let $\preceq$ be any well order satisfying $v \prec v'$ whenever $s(v) < s(v')$ (this exists since s takes finitely many values on vertices). Note that $S_{\preceq v}$ is obtained from $S_{\prec v}$ by coning off $S_{\prec v} \cap \partial \operatorname{star} v$. We claim that this intersection is precisely the boundary of $\mathrm{star}^{(h,s)\downarrow}_{Y_{p\le h}} v$ in $Y^{(h,s)\le(h,s)(v)}_{p\le h}$, which we will denote B, and which is homeomorphic to $\mathrm{lk}^{(h,s)\downarrow}_{Y_{p\le h}} v$ and hence $(k-1)$-connected by assumption. The inclusion $S_{\prec v} \cap \partial\,\mathrm{star}\, v \subseteq B$ is clear. Since $S_{\prec v} \cap \partial\,\mathrm{star}\, v$ is a full subcomplex of $\partial\,\mathrm{star}\, v$, it suffices for the converse to verify that any vertex w adjacent to v with $(h,s)(w) < (h,s)(v)$ lies in $S_{\prec v}$. If $h(w) < h(v)$ then $h(w) \le h(v) - \varepsilon \le r - \varepsilon \le q$, so $w \in Y_{p\le h \le q}$. Otherwise $s(w) < s(v)$ and thus $w \prec v$. □

11.2.2 Negative Properties

When trying to disprove finiteness properties using Morse theory, one often faces the following problem: suppose the Morse function is on a contractible space and the ascending links are always $(m-1)$-connected but infinitely often not m-connected. One would like to say that every ascending link that is not m-connected cones off at least one previously non-trivial m-sphere, and thus there are m-spheres that only get coned off arbitrarily late (and hence the filtration is not essentially m-connected). But this argument does not work in general because it is possible that the m-sphere one is coning off was actually already homotopically trivial in the superlevel set, and then one is actually *producing* an $(m+1)$-sphere. This second option can be excluded if one can make sure that no $(m+1)$-spheres are ever coned off (for example, if the whole contractible space is $(m+1)$-dimensional) and then the argument works. In general though, the difference between killing m-spheres and producing $(m+1)$-spheres is not visible locally and one has to take a more global view. The following will be useful in doing so.

Observation 11.7 *Let an $(m-1)$-connected affine cell complex X be equipped with a Morse function $(h,s)\colon X \to \mathbb{R} \times \mathbb{R}$ and assume that all ascending links are $(m-2)$-connected. Then the filtration $(X_{t\le h})_{t\in\mathbb{R}}$ is essentially $(m-1)$-connected, if and only if $X_{p\le h}$ is $(m-1)$-connected for some p, if and only if all $X_{p'\le h}$ are $(m-1)$-connected for all $p' \le p$ (for some p).*

Proof Since ascending links are $(m-2)$-connected, the Morse Lemma implies that for any $p < q$ the pairs $(X_{p\le h}, X_{q\le h})$ are $(m-1)$-connected, so in particular the map $\pi_k(X_{q\le h} \hookrightarrow X_{p\le h})$ is an isomorphism for $k < m-1$ and surjective for $k = m-1$. Now for these p and q, we see that this map induces the trivial map in homotopy up to dimension $m-1$ if and only if the homotopy groups vanish on $X_{p\le h}$. We conclude that the filtration is essentially $(m-1)$-connected if and only if $X_{p \le h}$ is $(m-1)$-connected for some $p \in \mathbb{Z}$, or equivalently all $p' \le p$. □

11.3 Thompson's Group and the Stein–Farley Complex

Thompson's group F has appeared over the past decades in a variety of situations, and has proved to have many strange and interesting properties. It was the first example of a torsion-free group of infinite cohomological dimension that is of type F_∞ [8]. Since it is of type F_∞, one can ask what its BNSR-invariants $\Sigma^m(F)$ are, for arbitrary m. The $m = 1$ case was answered by Bieri, Neumann and Strebel in [4], and the $m \ge 2$ case by Bieri, Geoghegan and Kochloukova in [5]. The main result of the present work is a recomputation of the $\Sigma^m(F)$, making use of a CAT(0) cube complex on which F acts freely, called the *Stein–Farley complex* X.

11.3.1 The Group

The fastest definition of F is via its standard infinite presentation,

$$F = \langle x_i, i \in \mathbb{N} \mid x_j x_i = x_i x_{j+1}, i < j \rangle.$$

One can also realize it as the group of orientation-preserving piecewise linear homeomorphisms of the interval $[0,1]$ with dyadic slopes and breakpoints. For our purposes, the most useful definition is in terms of *split-merge tree diagrams.*

A *split-merge tree diagram* (T_-/T_+) consists of a binary tree T_- of "splits" and a binary tree T_+ of "merges", such that T_- and T_+ have the same number of leaves. The leaves of T_- and of T_+ are naturally ordered left to right and we identify them. Two split-merge tree diagrams are *equivalent* if they can be transformed into each other via a sequence of reductions or expansions. A *reduction* is possible if T_- and T_+ contain terminal carets whose leaves coincide. The reduction consists of deleting both of these carets. An expansion is the inverse of a reduction. We denote the equivalence class of (T_-/T_+) by $[T_-/T_+]$.

These $[T_-/T_+]$ are the elements of F. The multiplication, say of elements $[T_-/T_+]$ and $[U_-/U_+]$, written $[T_-/T_+] \cdot [U_-/U_+]$, is defined as follows. First note that T_+ and U_- admit a binary tree S that contains them both, so using expansions we have $[T_-/T_+] = [\hat{T}_-/S]$ and

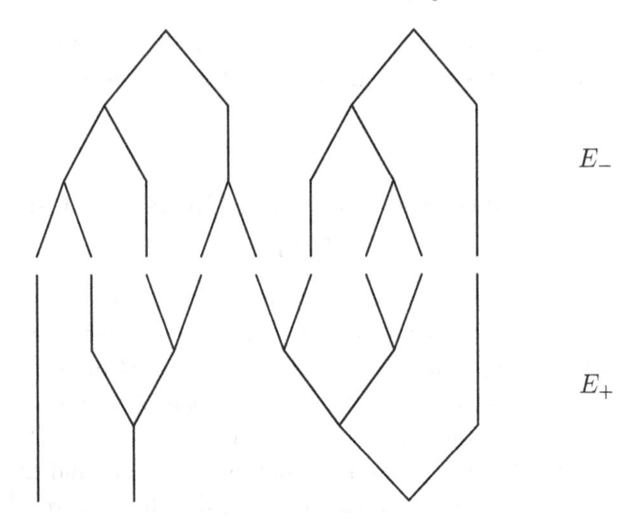

Figure 11.1 A split-merge diagram with two heads and three feet. The diagram is not reduced because the 7th and 8th leaves of E_- and E_+ both lie in terminal carets.

$[U_-/U_+] = [S/\hat{U}_+]$ for some $\hat{T}_-$ and $\hat{U}_+$. Then we can define

$$[T_-/T_+] \cdot [U_-/U_+] := [\hat{T}_-/S] \cdot [S/\hat{U}_+] = [\hat{T}_-/\hat{U}_+].$$

This multiplication is well defined, and it turns out that the resulting structure is a group, namely F. More information on the background of F can be found in [10]. We should point out that, under our convention, T_- encodes the subdivision of the range and T_+ of the domain; this is the reverse of the convention in [10].

11.3.2 The Stein–Farley Complex

We now recall the Stein–Farley cube complex X on which F acts. This was first constructed by Stein in [13], and shown to be CAT(0) by Farley [11]. We begin by generalizing split-merge tree diagrams to allow for *forests*: a *split-merge diagram* (E_-/E_+) consists of a binary forest E_- of "splits" and a binary forest E_+ of "merges" such that E_- and E_+ have the same number of leaves. By a *binary forest* we mean a finite sequence of rooted binary trees. Thus the leaves of E_- and of E_+ are naturally ordered left to right and we identify them. As in Figure 11.1 we will usually draw E_+ upside down. We call the roots of E_- *heads* and the roots of E_+ *feet* of the diagram.

Just like the tree case, we have a notion of equivalence using reduction and expansion, defined the same way. It is easy to see that every equivalence class contains a unique reduced split-merge diagram: just note that if a diagram (E_-/E_+) has two possible reductions that lead to diagrams (D^1_-/D^1_+) and (D^2_-/D^2_+), both reductions can be performed at once to give a diagram (C_-/C_+) that is a common reduction of (D^1_-/D^1_+) and (D^2_-/D^2_+) (i.e., reduction is confluent in a very strong way). We will sometimes abuse language and speak of a split-merge diagram, when really we are talking about equivalence classes of split-merge diagrams.

Let $\mathcal{P}$ be the set of equivalence classes of split-merge diagrams. This set has two important pieces of structure.

Groupoid. The first is a groupoid structure. If $[E_-/E_+]$ has k heads and ℓ feet, and $[D_-/D_+]$ has ℓ heads and m feet, then we will define their product, written $[E_-/E_+]\cdot[D_-/D_+]$, which is a split-merge diagram with k heads and m feet. Like in F, with split-merge tree diagrams, one way to define the product is to first note that E_+ and D_- admit a binary forest C that contains them both. Using expansions one can thus write $[E_-/E_+] = [\hat{E}_-/C]$ and $[D_-/D_+] = [C/\hat{D}_+]$ for some $\hat{E}_-$ and $\hat{D}_+$. Now we define

$$[E_-/E_+]\cdot[D_-/D_+] := [\hat{E}_-/C]\cdot[C/\hat{D}_+] = [\hat{E}_-/\hat{D}_+].$$

There is also a more visual description. The product can be obtained by stacking the diagram (E_-/E_+) on top of (D_-/D_+) and then applying a sequence of operations that are dual to expansion/reduction, namely if a merge is immediately followed by a split, both can be removed. Successively applying this operation eventually leads to a split-merge diagram that represents $[E_-/E_+]\cdot[D_-/D_+]$.

For this multiplication operation to give $\mathcal{P}$ a groupoid structure, we need identities and inverses. A forest in which all trees are trivial is called a *trivial forest.* The trivial forest with n trees is denoted id_n. We consider binary forests as split-merge diagrams via the embedding $E \mapsto [E/\mathrm{id}_m]$ where m is the number of leaves of E. In particular, for every m we have the element $[\mathrm{id}_m/\mathrm{id}_m]$, and this is clearly a multiplicative identity for split-merge diagrams against which it can be multiplied. Inverses are straightforward: the (left and right) inverse of $[E_-/E_+]$ is $[E_+/E_-]$.

Observation 11.8 *$\mathcal{P}$ is a groupoid with the above multiplication.*

Since F lives in $\mathcal{P}$ as the set of elements with one head and one foot, we have an action of F, by multiplication, on the subset $\mathcal{P}_1$ of elements with one head.

The second piece of structure on $\mathcal{P}$ is an order relation.

Poset. The order is defined by $[E_-/E_+] \leq [D_-/D_+]$ whenever there is a binary forest C such that $[E_-/E_+] \cdot C = [D_-/D_+]$ (recall that we identified the binary forest C with a split-merge diagram $[C/\mathrm{id}_m]$). In words, $[D_-/D_+]$ is greater than $[E_-/E_+]$ if it can be obtained from it by splitting feet. It is straightforward to check that $\leq$ is a partial order. The subset $\mathcal{P}_1$ of elements with one head is a subposet.

The topological realization of the poset $(\mathcal{P}_1, \leq)$ is a simplicial complex on which F acts, and the *Stein–Farley complex* X is a certain invariant subcomplex with a natural cubical structure. Given split-merge diagrams $[E_-/E_+] \leq [E_-/E_+] \cdot E$, we write $[E_-/E_+] \preceq [E_-/E_+] \cdot E$ if E is an *elementary forest.* This means that each of its trees is either trivial, or a single caret. Now X is defined to be the subcomplex of $|\mathcal{P}_1|$ consisting of those chains $x_0 < \cdots < x_k$ with $x_i \preceq x_j$ for all $i \leq j$. The cubical structure is given by intervals: given $x \preceq y$, the interval $[x, y] := \{z \mid x \leq z \leq y\}$ is a Boolean lattice of dimension n, and so the simplices in $[x, y]$ glue together into an n-cube. Here n is the number of carets in E, with $y = x \cdot E$.

Theorem 11.9 *[11] X is a* CAT(0) *cube complex.*

Note that the action of F on X is free. It is free on vertices since the action is just by multiplication in a groupoid. Also, it is free on cubes since if a group element stabilizes $[x, y]$ it must fix x and y.

Every cube σ has a unique vertex x with fewest feet and a unique vertex y with most feet. There is a unique elementary forest E with $y = x \cdot E$, and the other vertices of σ are obtained by multiplying x by subforests of E. We introduce some notation for this: suppose x has k feet and $E = (A_1, \dots, A_k)$, where each A_i is either I or Λ; here I is the trivial tree and Λ is the tree with one caret. Let Φ be the set of subforests of E, written $\Phi := \langle A_1, \dots, A_k \rangle$. Then the vertex set of σ is precisely $x\Phi$.

If we take a different vertex z of σ as "basepoint", then we also have to allow merges. Say z has $m > k$ feet. Then we can write $\sigma = z\Psi$ where Ψ is now of the form $\langle A_1, \dots, A_m \rangle$ where each A_i is either I, Λ or V. Here V is the inverse of the tree with one caret (so an upside-down caret). The tuple $(A_1, \dots, A_m)$ is now to be thought of as a split-merge diagram, and the set Ψ as the set of all split-merge diagrams that can be obtained by removing some of the carets. As before, the vertex set of σ is $z\Psi$.

11.3.3 Characters and Character Height Functions

It is well known that $\mathrm{Hom}(F, \mathbb{R}) \cong \mathbb{R}^2$. A standard choice of basis is $\{\chi_0, \chi_1\}$, where χ_0 and χ_1 are most easily described when viewing elements of F as piecewise linear homeomorphisms of $[0,1]$ with dyadic slopes and breakpoints. Then $\chi_0(f) := \log_2(f'(0))$ and $\chi_1(f) := \log_2(f'(1))$. Here the derivatives are taken on the right for χ_0 and the left for χ_1. So any character of F is of the form $\chi = a\chi_0 + b\chi_1$ for $a, b \in \mathbb{R}$.

For a tree T we define $L(T)$ to be the number of carets above the leftmost leaf of T, and $R(T)$ to be the number of carets above the rightmost leaf. Then the characters $\chi_i \colon F \to \mathbb{Z}$ can be expressed in terms of split-merge tree diagrams as

$$\chi_0([T_-/T_+]) := L(T_+) - L(T_-) \text{ and } \chi_1([T_-/T_+]) := R(T_+) - R(T_-). \tag{11.1}$$

It is readily checked that (11.1) is invariant under the equivalence relation on split-merge tree diagrams, and thus gives well defined maps. Replacing binary trees by binary forests, the above definition generalizes verbatim to arbitrary split-merge diagrams. In particular, the χ_i can now be evaluated on vertices of X. Moreover, any character χ on F can be written as a linear combination

$$\chi = a\chi_0 + b\chi_1 \tag{11.2}$$

and thus extends to arbitrary split-merge diagrams by interpreting (11.2) as a linear combination of the extended characters.

Since χ will be our height function, we need the following.

Lemma 11.10 *Any character χ extends to an affine map $\chi\colon X \to \mathbb{R}$.*

Proof It suffices to show that χ_0, χ_1 can be affinely extended. By symmetry it suffices to treat χ_0. Let $\square_2 = v\Phi$ be a square, say $\Phi = \langle A_1, \dots, A_k \rangle$, with exactly two A_i being Λ and all others being I. Say $A_i = A_j = \Lambda$ for $i < j$. Now either $i > 1$ and χ_0 is constant on $\square_2$, or $i = 1$ and χ_0 is affine-times-constant on $\square_2$. We conclude using Lemma 11.11 below. $\square$

Lemma 11.11 *A map $\varphi\colon \{0,1\}^n \to \mathbb{R}$ can be affinely extended to the cube $[0,1]^n$ if it can be affinely extended to its 2-faces.*

Proof The values of φ on the zero vector and on the standard basis vectors define a unique affine map $\tilde{\varphi}$. The goal is therefore to show that $\tilde{\varphi}$ coincides with φ on all the other vertices of $[0,1]^n$. This is proved for $v \in \{0,1\}^n$ by induction on the number of entries in v equal to 1. Let

$v = (v_i)_{1 \le i \le n}$ with $v_i = v_j = 1$ for some $i \neq j$. We know by induction that $\varphi(w) = \tilde{\varphi}(w)$ for the three vertices w obtained from v by setting to 0 the entries with index i or j or both. But these three vertices together with v span a 2-face, and so $\tilde{\varphi}(v)$ is the value of the (unique) affine extension of φ to that 2-face. Thus $\varphi(v) = \tilde{\varphi}(v)$. □

These extended characters χ will be our height functions. Our secondary height will be given by the number of feet function or its negative.

Observation 11.12 *There is a map $f\colon X \to \mathbb{R}$ that is affine on cubes and assigns to any vertex its number of feet.* □

Since our definition of Morse function required the secondary height function to take only finitely many values on vertices, for the next proposition we must restrict to subcomplexes of the form $X_{p \le f \le q}$. This is the full subcomplex supported on those vertices v with $p \le f(v) \le q$.

Proposition 11.13 *Let χ be a character. The pair (χ, f), as well as the pair $(\chi, -f)$, is a Morse function on $X_{p \le f \le q}$, for any $p \le q < \infty$.*

Proof We have already seen in Lemma 11.10 and Observation 11.12 that χ and f are affine. Also, f takes finitely many values on vertices in $X_{p \le f \le q}$. It remains to see that there is an $\varepsilon > 0$ such that any two adjacent vertices x and x' either have $|\chi(x) - \chi(x')| \ge \varepsilon$, or else $\chi(x) = \chi(x')$ and $f(x) \neq f(x')$. Let $\varepsilon = \min\{|c| \mid c \in \{a, b\} \setminus \{0\}\}$, where $\chi = a\chi_0 + b\chi_1$. We obtain x' from x by adding one split or one merge to the feet of x'. If it does not involve the first or last foot, then $\chi(x') = \chi(x)$. Otherwise $\chi(x') = \chi(x) \pm c$ for $c \in \{a, b\}$, and so $|\chi(x) - \chi(x')| = |c|$, which is either 0 or at least ε. □

11.4 Links and Subcomplexes

Since we will be doing Morse theory on X, we will need to understand homotopy properties of links in X. In this section we model the links, and then discuss some important subcomplexes of X.

11.4.1 (General) Matching Complexes

In this subsection we establish a useful model for vertex links in X.

Definition 11.14 Let Δ be a simplicial complex. A *general matching* is a subset μ of Δ such that any two simplices in μ are disjoint. The set of all general matchings, ordered by inclusion, is a simplicial complex, which we call the *general matching complex* $\mathcal{GM}(\Delta)$. For $k \in \mathbb{N}_0$ a *k-matching* is a general matching that consists only of k-cells. The set of all k-matchings forms the *k-matching complex*. If Δ is a graph, its 1-matching complex is the classical *matching complex*, denoted by $\mathcal{M}(\Delta)$.

By L_n we denote the linear graph on n vertices. Label the vertices $v_1 \ldots, v_n$ and the edges $e_{1,2}, \ldots, e_{n-1,n}$, so $e_{i,i+1}$ has v_i and v_{i+1} as endpoints $(1 \leq i \leq n-1)$.

Lemma 11.15 *$\mathcal{M}(L_n)$ is $(\lfloor \frac{n-2}{3} \rfloor - 1)$-connected.*

Proof For $n \geq 2$, $\mathcal{M}(L_n)$ is non-empty. Now let $n \geq 5$. Note that $\mathcal{M}(L_n) = \operatorname{star}(\{e_{n-2,n-1}\}) \cup \operatorname{star}(\{e_{n-1,n}\})$, a union of contractible spaces, with intersection $\operatorname{star}(\{e_{n-2,n-1}\}) \cap \operatorname{star}(\{e_{n-1,n}\}) \cong \mathcal{M}(L_{n-3})$. The result therefore follows from induction. □

It turns out that links of vertices in the Stein–Farley complex X are general matching complexes. Let x be a vertex of X with $f(x) = n$, where f is the "number of feet" function from Observation 11.12. The cofaces of x are precisely the cells $\sigma = x\Psi$, for every Ψ such that $x\Psi$ makes sense. In particular, if $\Psi = \langle A_1, \ldots, A_r \rangle$ for $A_i \in \{\mathrm{I}, \Lambda, \mathrm{V}\}$ $(1 \leq i \leq r)$, then the rule is that n must equal the number of A_i that are I or Λ, plus twice the number that are V.

Observation 11.16 *If a vertex x has $f(x) = n$ feet then* $\operatorname{lk} x \cong \mathcal{GM}(L_n)$.

Proof The correspondence identifies a simplex $x\Psi$, for $\Psi = \langle A_1, \ldots, A_r \rangle$, with a matching where (from left to right) I corresponds to a vertex not in the matching, Λ corresponds to a vertex in the matching and V corresponds to an edge in the matching. □

As a remark, under the identification $\operatorname{lk} x \cong \mathcal{GM}(L_n)$, the part of $\operatorname{lk} x$ corresponding to the matching complex $\mathcal{M}(L_n)$ is the *descending link with respect to f*. The higher connectivity properties of these descending links are crucial to proving that F is of type F_∞ using X (originally proved by Brown and Geoghegan [8] using a different space).

11.4.2 Restricting Number of Feet

Recall the subcomplex $X_{p\le f\le q}$ from Proposition 11.13. This is the full cubical subcomplex of X supported on the vertices x with $p \le f(x) \le q$. Similar notation is applied to define related complexes (e.g., where one inequality is strict or is missing).

Observation 11.17 *For $p, q \in \mathbb{N}$, the action of F on $X_{p\le f\le q}$ is co-compact.*

Proof For each n, F acts transitively on vertices with n feet. The result follows since X is locally compact. □

Lemma 11.18 *The complex $X_{p\le f\le q}$ is $\min(\lfloor\frac{q-1}{3}\rfloor - 1, q-p-1)$-connected. In particular, $X_{p\le f\le q}$ is $(\lfloor\frac{q-1}{3}\rfloor - 1)$-connected for any $p \le \lceil\frac{2q+1}{3}\rceil$.*

Proof We first show that $X_{p\le f}$ is contractible for every p. Since we know that X is contractible, it suffices to show that the ascending link with respect to f is contractible for every vertex x of X. We can then apply the Morse Lemma (using the Morse function $(f, 0)$). Indeed, the ascending link is an $(f(x)-1)$-simplex spanned by the cube $x\Phi$ where $\Phi = \langle\Lambda, \dots, \Lambda\rangle$. In particular, it is contractible.

Now we filter $X_{p\le f}$ by the spaces $X_{p\le f\le q}$, and so have to study descending links with respect to f. The descending link in X of a vertex x with $f(x) > q$ is isomorphic to $\mathcal{M}(L_{f(x)})$, which is $(\lfloor\frac{q-1}{3}\rfloor - 1)$-connected by Lemma 11.15. But in $X_{p\le f}$ only the $(f(x)-p-1)$-skeleton of that link is present. The descending link is therefore $\min(\lfloor\frac{q-1}{3}\rfloor - 1, q-p-1)$-connected. □

Corollary 11.19 *$X_{2k+1\le f\le 3k+1}$ is $(k-1)$-connected for every k.*

11.5 The Long Interval

We are now ready to compute $\Sigma^m(F)$ for all m, using the action of F on X. In this and the following two sections, we focus on different parts of the character sphere $S(F)$.

Let $\chi = a\chi_0 + b\chi_1$ be a non-trivial character of F. In this section we consider the case when $a < 0$ or $b < 0$. The corresponding part of $S(F) = S^1$ was termed the "long interval" in [5]. By symmetry we may assume $a \le b$ (so $a < 0$). We will show that for any $m \in \mathbb{N}$, $[\chi] \in \Sigma^m(F)$.

Let

$$n := 3m + 4.$$

Let $X_{2\leq f\leq n}$ be the sublevel set of X supported on vertices x with $2 \leq f(x) \leq n$. This is $(\lfloor \frac{n-1}{3} \rfloor - 1)$-connected, by Lemma 11.18, and hence is m-connected. It is also F-cocompact by Observation 11.17. Thus, by Corollary 11.3, to show that $[\chi] \in \Sigma^m(F)$, it suffices to show that the filtration $(X^{t\leq\chi}_{2\leq f\leq n})_{t\in\mathbb{R}}$ is essentially $(m-1)$-connected. To do this, we use the function

$$(\chi, -f)\colon X_{2\leq f\leq n} \to \mathbb{R} \times \mathbb{R}.$$

This is a Morse function by Proposition 11.13. By the Morse Lemma, the following lemma suffices to prove that in fact $X^{t\leq\chi}_{2\leq f\leq n}$ is $(m-1)$-connected for all $t \in \mathbb{R}$.

Lemma 11.20 *Let x be a vertex in $X_{2\leq f\leq n}$. Then the ascending link $\mathrm{lk}^{(\chi,-f)\uparrow}_{X_{2\leq f\leq n}} x$ in $X_{2\leq f\leq n}$ is $(m-1)$-connected.*

Proof Let $L := \mathrm{lk}^{(\chi,-f)\uparrow}_{X_{2\leq f\leq n}} x$. Vertices of L are obtained from x in two ways: either by adding a split to a foot or a merge to two adjacent feet. For such a vertex to be ascending, in the first case the split must strictly increase χ and in the second case the merge must not decrease χ. Also note that we only have simplices whose corresponding cubes lie in $X_{2\leq f\leq n}$. For instance if $f(x) = n$ then the vertices of L may only be obtained from x by adding merges.

We first consider the case when $f(x) = n$, so L is simply the subcomplex of $\mathcal{M}(L_n)$ consisting of matchings that do not decrease χ. Since $a < 0$ (and $n > 2$), merging the first and second feet decreases χ. Merging the $(n-1)$st and nth feet can either decrease, increase, or preserve χ, depending on whether $b < 0$, $b > 0$ or $b = 0$. Any other merging preserves χ and increases $-f$. Hence L is either $\mathcal{M}(L_{n-1})$ or $\mathcal{M}(L_{n-2})$, and in either case is $(\lfloor \frac{n-4}{3} \rfloor - 1)$-connected (Lemma 11.15) and hence $(m-1)$-connected.

The second case is when $f(x) < n-1$. Thinking of L as the subcomplex of $\mathcal{GM}(L_{f(x)})$ supported on vertices with $(\chi, -f)$-value larger than $(\chi, -f)(x)$, since $a < 0$ we know $\{e_{1,2}\} \notin L$. Also, the only vertices of $\mathcal{GM}(L_{f(x)})$ of the form $\{v_i\}$ that are in L are $\{v_1\}$ and possibly $\{v_{f(x)}\}$; since $f(x) < n-1$ this implies that for any $\mu \in L$ we have $\mu \cup \{v_1\} \in L$. Hence L is contractible with cone point $\{v_1\}$.

It remains to consider the case when $f(x) = n-1$. If $b \geq 0$, so $\{v_{n-1}\} \notin L$, then we can cone L off on $\{v_1\}$, as in the previous case.

Now suppose $b < 0$, so the vertices of L are $v_1, e_{2,3}, \dots, e_{n-3,n-2}, v_{n-1}$. Since $f(x) = n - 1$, v_1 and v_{n-1} do not span an edge in L, since this would involve $n + 1$ feet. Hence L is the join of $\{v_1\} \cup \{v_{n-1}\} \simeq S^0$ with $\mathcal{M}(L_{n-3})$. This is $\lfloor \frac{n-5}{3} \rfloor$-connected by Lemma 11.15, so is $(m-1)$-connected. □

The lemma together with the Morse Lemma gives:

Proposition 11.21 *If $a < 0$ or $b < 0$ then $[\chi] \in \Sigma^\infty(F)$.*

11.6 The Characters χ_0 and χ_1

Let $\chi = a\chi_0 + b\chi_1$ be a non-trivial character of F with $a > 0$ and $b = 0$, or $b > 0$ and $a = 0$. In this section we show that $[\chi] \notin \Sigma^1(F)$. We do this by considering the Morse function $(\chi, f)\colon X_{3 \le f \le 4} \to \mathbb{R} \times \mathbb{R}$. Thanks to Lemma 11.18 and Observation 11.17, $X_{3 \le f \le 4}$ is connected and F-cocompact. Hence by Corollary 11.3 it suffices to show that the filtration $(X^{t \le \chi}_{3 \le f \le 4})_{t \in \mathbb{R}}$ is not essentially connected. We would like to apply Observation 11.7, for which we need the following.

Lemma 11.22 *In $X_{3 \le f \le 4}$ all (χ, f)-ascending vertex links are non-empty.*

Proof Write $Y := X_{3 \le f \le 4}$. Up to scaling and symmetry we may assume that $\chi = \chi_0$. Given a vertex v on $n = f(v)$ feet, a simplex in $\mathrm{lk}_X\, v \cong \mathcal{GM}(L_n)$ is ascending if and only if its vertices are among $e_{1,2}$ (χ-ascending) and $v_2, \dots, v_n$ (f-ascending). The number of feet imposes restrictions on which of these simplices actually lie in $\mathrm{lk}_Y\, x$. But any ascending link contains $e_{1,2}$ or v_2 and thus is non-empty. □

Proposition 11.23 *If $a > 0$ and $b = 0$ or $b > 0$ and $a = 0$, then $[\chi] \notin \Sigma^1(F)$.*

Proof We treat the case $a > 0$, $b = 0$; the other case follows from symmetry. By positive scaling we can assume $a = 1$, so $\chi = \chi_0$. We want to show that $(X^{t \le \chi_0}_{3 \le f \le 4})_{t \in \mathbb{R}}$ is not essentially connected. By Observation 11.7 and Lemma 11.22 it suffices to show that $X^{t \le \chi_0}_{3 \le f \le 4}$ is not connected for any $t \in \mathbb{Z}$. Since F acts transitively on these sets, this is equivalent to proving that $X^{0 \le \chi_0}_{3 \le f \le 4}$ is not connected.

Let x be a vertex in X, and let (T/E) be its unique reduced representative diagram. Define $L(x) := L(T)$. Recall that $L(T)$ is the number of carets above the leftmost leaf of T. Let x' be any neighbor of

x, say with reduced diagram (T'/E'), so $L(x') := L(T')$. Note that $L(x') \in \{L(x)-1, L(x), L(x)+1\}$. We claim that if these neighboring vertices x, x' are in $X^{0\le\chi_0}_{3\le f\le 4}$, then $L(x) = L(x')$.

First note that $L(E) \ge L(x)$ because $\chi_0(x) \ge 0$. Also, $L(x) \ge 1$ since $f(x) \ge 3$, so also $L(E) \ge 1$. Now if x' is obtained from x by adding a merge to the feet of x, then T and T' have the same number of carets above the leftmost leaf, so $L(x) = L(x')$. Alternately, if x' is obtained from x by adding a split, since the leftmost leaf of E has at least one caret above it, we know that, again, adding this split cannot change the number of splits above the leftmost leaf of T. Hence in any case $L(x') = L(x)$ for all neighbors x' of x.

This shows that L is constant along the vertices of any connected component of $X^{0\le\chi_0}_{3\le f\le 4}$. Since L does take different values, there must be more than one component. □

11.7 The Short Interval

Let $\chi = a\chi_0 + b\chi_1$ be a non-trivial character of F, with $a > 0$ and $b > 0$. The corresponding part of $S(F) = S^1$ was termed the "short interval" in [5]. In this section we show that $[\chi] \in \Sigma^1(F) \setminus \Sigma^2(F)$. Consider the Morse function $(\chi, f)\colon X_{4\le f\le 7} \to \mathbb{R} \times \mathbb{R}$. By Lemma 11.18 and Observation 11.17, $X_{4\le f\le 7}$ is simply connected and F-cocompact. Hence by Corollary 11.3 it suffices to show that the filtration $(X^{t\le\chi}_{4\le f\le 7})_{t\in\mathbb{R}}$ is essentially connected but not essentially simply connected. We would like to apply Observation 11.7, for which we need the following.

Lemma 11.24 *In $X_{4\le f\le 7}$ all (χ, f)-ascending links are connected.*

Proof Given a vertex x with $n = f(x)$ feet, a simplex in $\mathrm{lk}_X\, x \cong \mathcal{GM}(L_n)$ is ascending if and only if all of its vertices lie in the set $\{e_{1,2}, e_{n-1,n}, v_2, \ldots, v_{n-1}\}$. The number of feet imposes restrictions on which of these simplices actually lie in $\mathrm{lk}_{X_{4\le f\le 7}}\, x$.

We claim that all ascending links are connected. For $n \in \{4,5\}$, v_2 through v_{n-1} are pairwise connected by an edge, and each of $e_{1,2}$ and $e_{n-1,n}$ is connected to at least one of them. For $n \in \{6,7\}$ the 0-simplices $e_{1,2}$ and $e_{n-1,n}$ are connected by an edge, and each v_i is connected to at least one of them. Thus in either case the ascending link is connected. □

To show that superlevel sets are not simply connected we will use the following supplement to the Nerve Lemma.

Lemma 11.25 *Let a simplicial complex Z be covered by connected subcomplexes $(Z_i)_{i\in I}$. Suppose the nerve $\mathcal{N}(\{Z_i\}_{i\in I})$ is connected but not simply connected. Then Z is connected but not simply connected.*

Proof That Z is connected follows for example from the usual Nerve Lemma, e.g., [6, Lemma 1.2]. However, the usual Nerve Lemma does not imply that Z is not simply connected, for example, since we are not assuming that the Z_i are simply connected. Hence, there is some work to do to prove that Z is not simply connected.

Let $N := \mathcal{N}(\{Z_i\}_{i\in I})$. For each $i \in I$ pick a vertex $z_i \in Z_i$. For any edge $\{i, j\}$ in N, pick an edge path $p_{i,j}$ from z_i to z_j in $Z_i \cup Z_j$ (this is possible because Z_i and Z_j are connected and meet non-trivially). This induces a homomorphism $\varphi\colon \pi_1(N, i) \to \pi_1(Z, z_i)$ where $i \in I$ is arbitrary. We want to see that φ is injective.

Let $\{i_0, i_1\}, \ldots, \{i_{n-1}, i_n\}, \{i_n, i_0\}$ be an edge cycle γ in N. Allow for the possibility of degenerate edges, i.e., edges $\{i, i\}$ that are actually vertices. Consider an arbitrary sequence of vertices $v_0, \ldots, v_n$ of Z with $v_j \in Z_{i_j}$ for each j, such that the edge $\{v_j, v_{j+1}\}$ exists in $Z_{i_j} \cup Z_{i_{j+1}}$ (with subscripts taken mod $n+1$); call the resulting edge cycle c. Up to introducing degenerate edges to γ, such a sequence always exists, and we also allow for the possibility of degenerate edges in c. The condition that $\{v_j, v_{j+1}\} \subseteq Z_{i_j} \cup Z_{i_{j+1}}$ for all j will be referred to as γ and c being *linked*. We assume that c can be filled by a triangulated disk in Z and want to show that γ is nullhomotopic in N.

The proof is by induction on the number of triangles in a (minimal) such filling disk. Let $t \subseteq Z$ be a triangle of a filling disk with vertices v_j, v_{j+1}, and w, say. Let k be such that Z_k contains t. Then we obtain a path homotopic to c by replacing the edge $\{v_j, v_{j+1}\}$ by the union of edges $\{v_j, w\}$ and $\{w, v_{j+1}\}$, and we obtain a path homotopic to γ by replacing the edge $\{i_j, i_{j+1}\}$ by the edges $\{i_j, k\}$ and $\{k, i_{j+1}\}$. Note that γ and c remain linked after this process, since $\{v_j, w\}$ and $\{w, v_{j+1}\}$ lie in Z_k. After finitely many such reductions, the filling disk for c contains no triangles.

To reduce c to a trivial cycle, the only remaining reductions needed are removing two forms of stuttering: the one where $v_j = v_{j+1}$ and the one where $v_j = v_{j+2}$. In the second case, we may remove v_{j+1} from c to obtain a homotopic path and we may similarly remove i_{j+1} from γ to obtain a (linked) homotopic path (note that i_j, i_{j+1} and i_{j+2} span a triangle because $Z_{i_j} \cap Z_{i_{j+1}} \cap Z_{i_{j+2}}$ contains either v_j or v_{j+1}, thanks to γ and c being linked). This reduces the second kind of stuttering to the first. The first kind of stuttering can be resolved if Z_{j+2} contains v_{j+1} by just deleting v_{j+1} and i_{j+1} from their respective paths. Otherwise (Z_{j+1} contains v_{j+2} and) the stuttering can be shifted by replacing v_{j+1} by v_{j+2} in c. Under any of these moves γ and c remain linked. After finitely many such reductions both c and γ will be trivial paths.

It follows that φ is injective: Let γ be a cycle in N and let c be the corresponding cycle in Z made up of the relevant paths $p_{k,\ell}$, as in the definition of φ. Each $p_{k,\ell}$ has a linked path consisting of $\{k,\ell\}$ and (possibly many occurrences of) $\{k\}$ and $\{\ell\}$. Thus γ and c are linked after sufficiently many degenerate edges have been added to γ. By the above argument, if $\varphi([\gamma]) = [c] \in \pi_1(Z, z_i)$ is trivial then so is $[\gamma] \in \pi_1(N, i)$. □

Proposition 11.26 *If $a > 0$ and $b > 0$ then $[\chi] \in \Sigma^1(F) \setminus \Sigma^2(F)$.*

Proof We want to show that the filtration $(X^{t\leq\chi}_{4\leq f\leq 7})_{t\in\mathbb{R}}$ is essentially connected but not essentially simply connected. First note that the Morse Lemma together with Lemma 11.24 shows that in fact every $X^{t\leq\chi}_{4\leq f\leq 7}$ is already connected.

To show that $(X^{t\leq\chi}_{4\leq f\leq 7})_{t\in\mathbb{R}}$ is not essentially simply connected, it suffices by Observation 11.7 and Lemma 11.24 to show that $X^{t\leq\chi}_{4\leq f\leq 7}$ fails to be simply connected, for arbitrarily small $t \in \mathbb{R}$. Since the action of F on $\mathbb{R}$ induced by χ is cocompact, it suffices to show that $X^{0\leq\chi}_{4\leq f\leq 7}$ is not simply connected.

To this end we consider certain subcomplexes of $Y := X^{0\leq\chi}_{4\leq f\leq 7}$. To define them, as in the proof of Proposition 11.23, for a vertex x of X with reduced representative (T/E), define $L(x) := L(T)$. Similarly let $R(x) := R(T)$. Now we consider the full subcomplexes $Y_{L=i}$ of Y supported on vertices x with $L(x) = i$. Similarly we have complexes $Y_{R=i}$. Each of these spaces decomposes into countably many connected components, which we enumerate as $Y^m_{B=i}$, $m \in \mathbb{N}$ (where $B = L$ or $B = R$).

We claim that Y is covered by the complexes $Y^m_{L=i}, m \in \mathbb{N}$ and $Y^m_{R=i}, m \in \mathbb{N}$. We have to show that every cell $x\Psi$ in Y is contained either in some $Y_{L=i}$ or some $Y_{R=i}$. Take x to have the maximal number of feet within the cell so that Ψ involves only I and V. Note that in (T/E), which is the reduced representative of x, we know that T is nontrivial since $f(x) \geq 4$, and so $L(T) > 0$ and $R(T) > 0$. Hence, in order for $\chi(x) \geq 0$ to hold we need at least one of $L(E) > 0$ or $R(E) > 0$ to hold. If $L(E) > 0$, then $L(y) = L(x)$ for all vertices y of $x\Psi$ (for similar reasons as in the proof of Proposition 11.23), with a similar statement for the other case, and hence $x\Psi$ is contained in some $Y_{B=i}$.

Note that by definition $Y_{B=i} \cap Y_{B=j} = \emptyset$ for $i \neq j$ and $B \in \{L, R\}$. Thus the nerve $\mathcal{N}(\{Y^m_{B=i}\}_{i,m\in\mathbb{N},B\in\{L,R\}})$ is a bipartite graph. That it is connected can be deduced from [6, Lemma 1.2] since Y is connected.

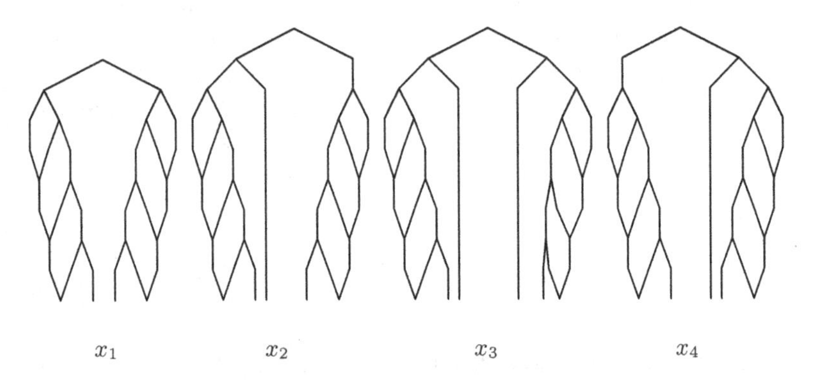

Figure 11.2 The diagrams used in the proof of Proposition 11.26.

To construct an explicit cycle, consider the vertices x_1 to x_4 in Figure 11.2. Note that x_4 and x_1 lie in the same component of $Y_{L=2}$. This is because after extending the left hand side by splitting the second foot and then merging the first two feet sufficiently many times (depending on a and b), χ_0 is so high that the right side can be removed without χ dropping below zero. Let us say that this component is $Y^0_{L=2}$. For similar reasons x_1 and x_2 lie in a common component $Y^0_{R=2}$, the vertices x_2 and x_3 lie in a common component $Y^0_{L=3}$ and x_3 and x_4 lie in a common component $Y^0_{R=3}$. It follows that $\mathcal{N}(\{Y^m_{B=i}\}_{i,m\in\mathbb{N},B\in\{L,R\}})$ contains the cycle $Y^0_{L=2}, Y^0_{R=2}, Y^0_{L=3}, Y^0_{R=3}$. Now Lemma 11.25 tells us that Y is not simply connected (it applies to simplicial complexes, so we take a simplicial subdivision). □

Acknowledgments

Thanks are due to Robert Bieri for suggesting this problem to us, and to Matt Brin for helpful discussions. We also thank the anonymous referee for useful comments and suggestions.

The first author gratefully acknowledges support by the DFG through the project WI 4079/2 and through the SFB 701.

References

[1] Bestvina, Mladen and Brady, Noel. 1997. Morse theory and finiteness properties of groups. *Invent. Math.*, **129**(3), 445–470.

[2] Bieri, Robert and Geoghegan, Ross. 2003. Connectivity properties of group actions on non-positively curved spaces. *Mem. Amer. Math. Soc.*, **161**(765), xiv+83.

[3] Bieri, Robert and Renz, Burkhardt. 1988. Valuations on free resolutions and higher geometric invariants of groups. *Comment. Math. Helv.*, **63**(3), 464–497.

[4] Bieri, Robert, Neumann, Walter D. and Strebel, Ralph. 1987. A geometric invariant of discrete groups. *Invent. Math.*, **90**(3), 451–477.

[5] Bieri, Robert, Geoghegan, Ross and Kochloukova, Dessislava H. 2010. The Sigma invariants of Thompson's group F. *Groups Geom. Dyn.*, **4**(2), 263–273.

[6] Björner, A., Lovász, L., Vrećica, S. T. and Živaljević, R. T. 1994. Chessboard complexes and matching complexes. *J. London Math. Soc. (2)*, **49**(1), 25–39.

[7] Bridson, Martin R. and Haefliger, André. 1999. *Metric spaces of non-positive curvature.* Grundlehren der Mathematischen Wissenschaften [Fundamental Principles of Mathematical Sciences], vol. 319. Springer-Verlag.

[8] Brown, Kenneth S. and Geoghegan, Ross. 1984. An infinite-dimensional torsion-free FP_∞ group. *Invent. Math.*, **77**(2), 367–381.

[9] Bux, Kai-Uwe. 2004. Finiteness properties of soluble arithmetic groups over global function fields. *Geom. Topol.*, **8**, 611–644.

[10] Cannon, J. W., Floyd, W. J. and Parry, W. R. 1996. Introductory notes on Richard Thompson's groups. *Enseign. Math. (2)*, **42**(3-4), 215–256.

[11] Farley, Daniel S. 2003. Finiteness and CAT(0) properties of diagram groups. *Topology*, **42**(5), 1065–1082.

[12] Geoghegan, Ross. 2008. *Topological methods in group theory.* Graduate Texts in Mathematics, vol. 243. Springer, New York.

[13] Stein, Melanie. 1992. Groups of piecewise linear homeomorphisms. *Trans. Amer. Math. Soc.*, **332**(2), 477–514.

Printed in the United States
By Bookmasters